Problem Solving Methods and Strategies in High School Mathematical Competitions

Mathematical Olympiad Series

ISSN: 1793-8570

Series Editors: Lee Peng Yee *(Nanyang Technological University, Singapore)*
Xiong Bin *(East China Normal University, China)*

Published

Vol. 21 *Problem Solving Methods and Strategies in High School*
Mathematical Competitions
edited by Bin Xiong (East China Normal University, China) &
Yijie He (East China Normal University, China)

Vol. 20 *Hungarian Mathematical Olympiad (1964–1997):*
Problems and Solutions
by Fusheng Leng (Academia Sinica, China),
Xin Li (Babeltime Inc., USA) &
Huawei Zhu (Shenzhen Middle School, China)

Vol. 19 *Mathematical Olympiad in China (2019–2020):*
Problems and Solutions
edited by Bin Xiong (East China Normal University, China)

Vol. 18 *Mathematical Olympiad in China (2017–2018):*
Problems and Solutions
edited by Bin Xiong (East China Normal University, China)

Vol. 17 *Mathematical Olympiad in China (2015–2016):*
Problems and Solutions
edited by Bin Xiong (East China Normal University, China)

Vol. 16 *Sequences and Mathematical Induction:*
In Mathematical Olympiad and Competitions
Second Edition
by Zhigang Feng (Shanghai Senior High School, China)
translated by: Feng Ma & Youren Wang

Vol. 15 *Mathematical Olympiad in China (2011–2014):*
Problems and Solutions
edited by Bin Xiong (East China Normal University, China) &
Peng Yee Lee (Nanyang Technological University, Singapore)

The complete list of the published volumes in the series can be found at
http://www.worldscientific.com/series/mos

Vol. 21 | Mathematical
Olympiad
Series

Problem Solving Methods and Strategies in High School Mathematical Competitions

Original Authors

Bin Xiong *East China Normal University, China*
Yijie He *East China Normal University, China*

English Translator

Yongming Liu *East China Normal University, China*

Copy Editors

Ming Ni *East China Normal University Press, China*
Lingzhi Kong *East China Normal University Press, China*
Yuanlin Wan *East China Normal University Press, China*

 **East China Normal
University Press**

 **World Scientific**

Published by

East China Normal University Press
3663 North Zhongshan Road
Shanghai 200062
China

and

World Scientific Publishing Co. Pte. Ltd.
5 Toh Tuck Link, Singapore 596224
USA office: 27 Warren Street, Suite 401-402, Hackensack, NJ 07601
UK office: 57 Shelton Street, Covent Garden, London WC2H 9HE

Library of Congress Cataloging-in-Publication Data
Names: Xiong, Bin, author. | He, Yijie, author.
Title: Problem solving methods and strategies in high school mathematical competitions / Bin Xiong
    (East China Normal University, China), Yijie He (East China Normal University, China) ;
    translated by Yongming Liu (East China Normal University, China).
Other titles: Gao zhong shu xue jing sai zhong de jie ti fang fa yu ce lue. English
Description: New Jersey : World Scientific, [2023] |
    Series: Mathematical Olympiad series, 1793-8570 ; Vol. 21
Identifiers: LCCN 2023030352 | ISBN 9789811277429 (hardcover) |
    ISBN 9789811278686 (paperback) | ISBN 9789811277436 (ebook for institutions) |
    ISBN 9789811277443 (ebook for individuals)
Subjects: LCSH: Mathematics--Problems, exercises, etc. | Mathematics--Competitions.
Classification: LCC QA43 .X5613 2023 | DDC 510.76--dc23/eng/20230817
LC record available at https://lccn.loc.gov/2023030352

British Library Cataloguing-in-Publication Data
A catalogue record for this book is available from the British Library.

For any available supplementary material, please visit
https://www.worldscientific.com/worldscibooks/10.1142/13442#t=suppl

Typeset by Stallion Press
Email: enquiries@stallionpress.com

Printed in Singapore

Preface for the Series of Books for Mathematical Olympiad

Mathematics competitions, like other competitive activities, are a kind of intellectual competitions for young students. In similar intellectual competitive activities with basic science as the content, mathematics competitions have the longest history, with worldwide profound influence. China began to hold mathematics competitions in 1956. At that time, the most esteemed, famous mathematicians Hua Luogeng, Su Buqing, Jiang Zehan all took an active part in the leadership and organization of the competition activities, and organized the publication of a series of mathematics books for young people, which inspired a large number of young students determined to engage in scientific careers. China has participated in the International Mathematical Olympiad since 1986, and won the first place in the overall score of the team many times. In 1990, China successfully held the 31st International Mathematical Olympiad in Beijing, which indicated that China's mathematical competition level was in the leading position in the world, attracting the attention of scientists and educators all over the world.

It is indicated that mathematics competitions, especially in the region and units where they are well organized and developed, can greatly stimulate students' interest in studying mathematics, help to cultivate creating thinking, and enhance students' learning efficiency.

This competitive activity, introducing a healthy competitive mechanism into the mathematics teaching process, is conducive to the selection of talented candidates. The winners selected by the mathematical competition not only have a solid and extensive mathematical foundation, but also have a diligent and scientific learning method. Many of the young students will become excellent scientific researchers in the future. In the United States, winners of mathematical competitions, such as J.W. Milnor, D.B. Mumford, D. Quillen, etc., were all winners of the Fields Prize. In Poland, the famous number theorist A. Schinzel was a math contest winner when he was a student. In Hungary, famous mathematicians L. Fejér, M. Riesz, G. Szegö, A. Haar and T. Radó were all winners of mathematics competitions. Hungary was the first country to hold mathematical competitions and has produced many great mathematicians out of proportion to its population!

While carrying out the activities of mathematical competitions, schools can strengthen ties and exchange experiences in teaching mathematics with each other. In this sense, mathematical competitions may become 'catalysts' for reform of mathematics curriculums and a powerful measure to cultivate excellent talents.

However, attention should be paid to the combination of popularization of and improvement in mathematical competitions, and priority should be given to popularization so that the competitions have a broad mass basis.

Of course, nowadays, some people pay too much attention to the results of mathematical contests. The organization and participation have a strong utilitarian purpose and expand the role of mathematical contests excessively. All these are incorrect and against the original intention of carrying out mathematical contests. These shortcomings have deep social causes and need to be overcome step by step. There is no need to relinguish this activity just because there are some shortcomings.

I am very happy to see the official publication of the Mathematical Olympiad Series. This set of books is large in scale and detailed in topics. As far as I know, such series are rare. This set of books not only elaborates the common methods to solve the problems of mathematics competitions, but also precisely analyszes them to determine the most appropriate, many of which directly come from the author's own research results. They may be considered not only as very good mathematics competitions subject course ware, but also as the primary and secondary school students' and teachers' reference books.

The authors are mathematical competition teachers and researchers, and many are national team coaches and national team leaders. They have made contributions to the development of mathematical competitions in China, to the achievements of Chinese students in IMO and the glory of China, and they have made painstaking efforts for the early release of this series of books. The Press of East China Normal University planned and organized this series of books on the basis of publishing (*Mathematical Olympiad Course*) *Problems and Solutions in Mathematical Olympiad*, *Towards IMO*, and other competition books. I am very grateful to the authors and editors for their work in this regard, and I sincerely wish the mathematics competitions in our country become better and better.

Wang Yuan

Contents

Chapter 1

Methods of Reduction

'Reduction method' in mathematics is a strategy of solving a problem by changing the problem into a simpler or solved problem by some transformations. In some sense, reduction is simplification. The famous Hungarian mathematician Rózsa Péter recorded a well-known joke in mathematical circles in her masterwork *Playing with Infinity: Mathematical Explorations and Excursions*, to illustrate how mathematicians solve problems by reduction.

'There is a gas ring in front of you, a tap, a saucepan, and a match. You want to boil some water. What do you do?' The reply is usually given with an air of uncertainty: 'I light the gas, put some water in the saucepan and put it in the gas.' 'So far you are quite correct. Now I shall modify the problem: everything is the same as before, the only difference is that there is enough water in the saucepan. Now what do you do?' Now the problem-solver speaks up, surer of himself, knowing himself to be in the right: 'I light the gas and put the saucepan on.' Then comes the superior reply: 'Only a physicist would do that. A mathematician would pour the water away and say that he had reduced the problem to the previous one!'

'Pour the water away!', that is reduction, the typical way that mathematicians think.

We begin with two examples as follows:

Example 1. Suppose that a sequence $\{a_n\}$ satisfies $a_1 = 1/3$, $a_{n+1} = \sqrt{(1 + a_n)/2}$, $n \in \mathbb{N}^+$. Find the expression of the general term a_n.

Solution. It is natural to make the trigonometric substitutions: $a_n = \cos b_n$ and $a_{n+1} = \cos b_{n+1} = \cos(b_n/2)$. This yields a geometric sequence

$\{b_n\}$, $b_{n+1} = b_n/2$, $b_1 = \arccos(1/3)$. The general term is $b_n = b_1/2^{n-1}$. Thus,

$$a_n = \cos b_n = \cos(\arccos(1/3)/2^{n-1}).$$

Example 2. If an equation of z: $z^2 + c|z|^2 = 1 + 2\mathrm{i}$ has complex solution(s), find the range of the real number c.

Solution. Let a complex solution be $z = a + b\mathrm{i}$, $a, b \in \mathbb{R}$. Then

$$a^2 - b^2 + 2ab\mathrm{i} + c(a^2 + b^2) = 1 + 2\mathrm{i}.$$

Since the real parts and the imaginary parts on both sides are equal, respectively, we have

$$\begin{cases} (c+1)a^2 + (c-1)b^2 = 1, \\ 2ab = 2. \end{cases}$$

Substitute $b = 1/a$ into the first equation to get

$$(c+1)a^4 - a^2 + (c-1) = 0.$$

where c can be solved as

$$c = \frac{1 + a^2 - a^4}{1 + a^4} = -1 + \frac{1}{a^2 + 2 + \frac{5}{a^2+2} - 4}.$$

By using the basic inequality $x^2 + y^2 \geq 2xy$, we have $-1 < c \leq -1 + 1/(2\sqrt{5} - 4) = \sqrt{5}/2$, and equality holds when $a^2 + 2 = \sqrt{5}$. Therefore, the range of c is $(-1, \sqrt{5}/2]$.

Remark. The solutions of Examples 1 and 2 use reduction method to change the original problems to simpler ones.

We give more examples to demonstrate the strategy of reduction.

Example 3. (1) Thirteen kids are standing in a circle. How many kids can be selected at most, such that any two of them are not neighbours in the circle?

(2) Among integers from 1 to 13, how many can be selected at most, such that the difference between any two of them is neither 5 nor 8.

Solution of (1). Label the kids by integers from 1 to 13 as Fig. 1.1 shows. Then at least 6 kids can be selected, for example, 1, 3, 5, 7, 9, and 11. Now we show that no more kid can be selected.

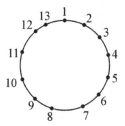

Figure 1.1

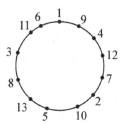

Figure 1.2

We first select one kid: without loss of generality, suppose it is labeled 1. Then there are five pairs of kids: (3,4), (5,6), (7,8), (9,10), and (11,12), from which at most one can be selected in each pair. That is to say, at most six kids can be selected.

Solution of (2). We reduce the problem (2) to (1).

We arrange integers 1 to 13, as shown in Fig. 1.2, such that two numbers are neighbouring if and only if their difference is 5 or 8. So, problem (2) is reduced to problem (1), and the answer is 6. For example, we can select 1, 4, 7, 10, 13, and 3.

Remark. We solved the difficult problem (2) by reducing it to a simpler problem (1).

Example 4. Find all sequences $\{x_n\}$ of positive integers satisfying the following two conditions:

(1) For each $n \in \mathbb{N}^+$, $x_n \leq n\sqrt{n}$;
(2) For any distinct $m, n \in \mathbb{N}^+$, $(n - m)\,|\,(x_n - x_m)$.

Solution. By (1), we see that $x_1 \le 1, x_2 \le 2\sqrt{2}$, and hence $x_1 = 1, x_2 = 1$ or 2.

Case 1. $x_1 = 1, x_2 = 1, x_n \ge 2 - n, n \in \mathbb{N}^+$. By (2), for $n > 2$, $(n - 1) \,|\, (x_n - 1)$ and $(n - 2) \,|\, (x_n - 1)$. Since $(n - 1, n - 2) = 1$ for $n > 2$, we deduce that $(n - 1)(n - 2) \,|\, (x_n - 1)$. This gives the inequations

$$2 - n \le x_n = 1 + c_n(n - 1)(n - 2) \le n\sqrt{n}, \quad c_n \in \mathbb{Z},$$

and we must have $c_n = 0$ for $n \ge 5$, namely, $x_n = 1$ for $n \ge 5$. Similarly, the inequations $-1 \le x_3 = 1 + a_3(n - 3) \le 3\sqrt{3}$ and $-2 \le x_4 = 1 + a_4(n - 4) \le 4$, $a_3, a_4 \in \mathbb{Z}$, for $n \ge 8$ give $a_3, a_4 = 0$, that is, $x_3 = 1, x_4 = 1$. So, for case (1), the sequence $\{x_n\}$ is $x_n = 1$ for $n \in \mathbb{N}^+$.

Case 2. $x_1 = 1, x_2 = 2$. Let $x_n' = x_n - (n - 1)$, then $x_n' \ge 2 - n$, $x_1' = 1 = x_2'$, and $\{x_n'\}$ satisfies (1), (2). By the result of case (1), $x_n' = 1$, That is $x_n = n$, $n \in \mathbb{N}^+$.

Remark. We can also solve for case 2 directly. For $n \ge 3$, there exist $a_n, b_n \in \mathbb{Z}$, such that $x_n = 1 + a_n(n - 1) = 2 + b_n(n - 2)$. Then $a_n = 1 + (b_n - a_n)(n - 2)$. Thus, $1 \le x_n = n + (b_n - a_n)(n - 1)(n - 2) \le n\sqrt{n}$. We obtain, $b_n - a_n = 0$ for $n \ge 4$, namely $x_n = n$ for $n \ge 4$.

At last, we solve inequations $1 \le x_3 = n + c_n(n - 3) \le 3\sqrt{3}$, $c_n \in \mathbb{Z}$ for $n \ge 6$, and obtain $c_n = -1$, that is, $x_3 = 3$. So, for case (2), the sequence $\{x_n\}$ is, $x_n = n$, $n \in \mathbb{N}^+$.

Example 5. Let $x_1, x_2, \ldots, x_n \in [0, 1]$, $n \ge 2$. Show that

$$n + \sum_{j=1}^{n} x_j^2 x_{j+1} \ge 2 \sum_{j=1}^{n} x_j^3,$$

where $x_{n+1} = x_1$.

Solution. First, we observe that a simple inequality $1 + x^2 y \ge x^3 + y^3$, $x, y \in [0, 1]$. It holds because

$$1 + x^2 y - x^3 - y^3 = \begin{cases} (1 - x^3) + y(x^2 - y^2) \ge 0, & \text{if } x \ge y; \\ (1 - y^3) + x^2(y - x) \ge 0, & \text{if } x \le y. \end{cases}$$

Then, the desire inequality follows from $\sum_{i=1}^{n} (1 + x_i^2 x_{i+1}) \ge \sum_{i=1}^{n} (x_i^3 + x_{i+1}^3)$.

Remark. (1) The inequality $1 + x^2 y \ge x^3 + y^3$, for $x, y \in [0, 1]$ can also be proved by considering the function $f(y) = y^3 - x^2 y - (1 - x^3)$ is convex down on the interval $[0, 1]$ with $f(0) \le 0$ and $f(1) \le 0$.

(2) For methods of reduction, sometimes a problem is transformed into an equivalent problem; sometimes it is transformed into a nonequivalent problem, but the new problem easily gives the solution of the original one. This kind of reduction is commonly used in the proof of inequalities.

Example 6. Let a, b, and c be the side lengths of a triangle. Show the following:

(i) $(b + c)(c + a)(a + b) \geq 8abc$;
(ii) $(b + c - a)(c + a - b)(a + b - c) \leq abc$;
(iii) $(b + c)(c + a)(a + b)(b + c - a)(c + a - b)(a + b - c) \leq 8a^2b^2c^2$.

Solution. (i) Clearly, $x + y \geq 2\sqrt{xy}$, $x, y \geq 0$. We have

$$(b + c)(c + a)(a + b) \geq 8\sqrt{bc}\sqrt{ca}\sqrt{ab} = 8abc.$$

(In fact, the inequality holds for any non-negative numbers a, b, c.)

(ii) By the properties of triangles, let $b+c-a = 2x > 0$, $c+a-b = 2y > 0$, and $a + b - c = 2z > 0$. Then we have $y + z = a, z + x = b, x + y = c$, and by (i), we have

$$(b + c - a)(c + a - b)(a + b - c) = 8xyz \leq (y + z)(z + x)(x + y) = abc.$$

(iii) Using symbols as in (ii). We write inequality (iii) as

$$(2x + y + z)(2y + z + x)(2z + x + y)xyz \leq (y + z)^2(z + x)^2(x + y)^2. \quad (3)$$

Note that

$$xy(2z + x + y)^2 = xy(x + y)^2 + 4xyz(x + y + z)$$
$$\leq xy(x + y)^2 + (x + y)^2 z(x + y + z)$$
$$= (x + y)^2(xy + z(x + y + z))$$
$$= (x + y)^2(z + x)(y + z).$$

That is

$$xy(2z + x + y)^2 \leq (x + y)^2(z + x)(y + z). \quad (4)$$

Similarly,

$$yz(2x + y + z)^2 \leq (y + z)^2(x + y)(z + x), \quad (5)$$

and

$$zx(2y + z + x)^2 \leq (z + x)^2(y + z)(x + y). \quad (6)$$

Multiplying the above three inequalities (4), (5), (6), and taking the square root, we obtain (3).

Remark. The substitutions in (ii) are often used to prove inequalities involving side lengths of a triangle. Inequality (ii) can also be proved directly by

$$(b + c - a)(a + b - c) = b^2 - (c - a)^2 \leq b^2$$

and analogous inequalities

$$(c + a - b)(b + c - a) \leq c^2,$$

$$(a + b - c)(c + a - b) \leq a^2.$$

Multiply the above three inequalities and take the square root to obtain (ii). Similar techniques can be used for proving (iii).

Note that (ii) can be obtained by (i) and (iii).

We point out that inequality (iii) holds for any positive real numbers a, b, and c. Since (ii) can be obtained by (i) and (iii), it holds for any positive real numbers a, b, and c as well. Please have a try.

Example 7. Let $b > a > 0$. Show that

$$\frac{\log b - \log a}{b - a} < \frac{1}{\sqrt{ab}}.$$

Solution. The original inequality is equivalent to $\log(b/a) < \sqrt{b/a} - \sqrt{a/b}$. Let $b/a = t^2, t > 1$. Then we are to show $F(t) = 2\log t - (t - 1/t) < 0$, when $t > 1$. This is clearly verified by $F(1) = 0$, and $F'(t) = -(1 - 1/t)^2 < 0$.

Remark. The above example comes from the 2002 China National Entrance Exam for Masters. The reduction method is used in several steps. The idea of reduction is described by the Hungarian American mathematician Pólya György in his famous book *How to Solve It, A New Aspect of Mathematical Method as follows*: 'We often have to try various modifications to the problem. We have to vary, to restate, to transform it again and again till we succeed eventually in finding something useful.'

Example 8. There are 20 balls in a circle labeled $1, 2, 3, \ldots, 20$ clockwise at the beginning. Define a *reversion* to be reverse the order of four successive balls (for example, 20, 1, 2, 3 to be 3, 2, 1, 20). Can you make some *reversions* to change the original order into the following orders?

(a) 5, 1, 2, 3, 4, 6, 7,..., 20;
(b) 6, 1, 2, 3, 4, 5, 7,..., 20.

Solution. (a) The answer is positive. Take four *reversions* as follows: $12345 \rightarrow 15432 \rightarrow 34512 \rightarrow 32154 \rightarrow 51234$. The positions of the other balls are unchanged during this process.

So, every ball can be shifted four places anticlockwise by four *reversions*, leaving all the other balls' orders unchanged.

(b) By taking four *reversions* in each step, we can move number 6 anticlockwise between 1 and 2, then between 17 and 18, then between 13 and 14, then between 9 and 10, then between 4 and 5, then at last between 20 and 1. So we can use $4 \times 6 = 24$ *reversions* to reach (b).

Example 9. Let $F(a,b,c) = \max_{x \in [1,3]} \left| x^3 - ax^2 - bx - c \right|$. When a, b, c run over all the real numbers, find the smallest possible value of F (Problem of 2011 China Team Selection Test).

Solution. Let

$$f(x) = (x+2)^3 - a(x+2)^2 - b(x+2) - c \qquad \cdot$$
$$= x^3 + (6-a)x^2 + (12 - 4a - b)x + (8 - 4a - 2b - c).$$

Then the problem is to find

$$\min_{a_1, b_1, c_1 \in \mathbb{R}} \max_{x \in [-1,1]} \left| x^3 + a_1 x^2 + b_1 x + c_1 \right|,$$

where $a_1 = 6 - a$, $b_1 = 12 - 4a - b$, $c_1 = 8 - 4a - 2b - c$ can be any real numbers since $a = 6 - a_1$ $b = 4a_1 - b_1 - 12$, $c = 8 - 4a_1 + 2b_1 - c_1$ are arbitrary.

Firstly, we show that $F = \max_{x \in [-1,1]} |f(x)| \geq 1/4$. Take $x = \pm 1, \pm 1/2$, and we have

$$F \geq f(1) = 1 + a_1 + b_1 + c_1,$$
$$F \geq -f(-1) = 1 - a_1 + b_1 - c_1,$$
$$F \geq -f(1/2) = -\frac{1}{8} - \frac{a_1}{4} - \frac{b_1}{2} - c_1,$$
$$F \geq f(-1/2) = -\frac{1}{8} + \frac{a_1}{4} - \frac{b_1}{2} + c_1.$$

Thus, $6F \geq f(1) - f(-1) - 2f(1/2) + 2f(-1/2) = \frac{3}{2}$, that is, $F \geq 1/4$.

On the other hand, if $F = 1/4$, then $f(1) = -f(-1) = -f(1/2) = f(-1/2) = 1/4$, which give $a_1 = c_1 = 0, b_1 = -3/4$. So, indeed we have $\min F = 1/4$.

Remark. By changing the range of x to a symmetric interval $[-1,1]$ about the origin, it is easier to evaluate the function and simplify the inequalities.

Exercises

1. Find the maximum of $x\sqrt{1-y^2}+y\sqrt{1-x^2}$ (Problem of 1990 Moscow Mathematical Olympiad).

2. Let $b > a > e = 2.718281828\ldots$. Show that $b^a < a^b$.

3. The number 5 can be written as the ordered sum of three positive integers in six different ways. They are: $5 = 1+1+3 = 1+3+1 = 3+1+1 = 1+2+2 = 2+1+2 = 2+2+1$.

   Let $m \leq n$ be two positive integers. In how many ways can n be written as the ordered sum of m positive integers?

4. For real numbers $x, y, z \geq 1$, show that

   $$(x^2 - 2x + 2)(y^2 - 2y + 2)(z^2 - 2z + 2) \leq (xyz)^2 - 2xyz + 2.$$

   (Problem from the 2009 China Girls Mathematical Olympiad)

5. Let a be the arithmetic mean of n real numbers $x_1, x_2, \ldots, x_n$. Show that

   $$\sum_{k=1}^{n}(x_k - a)^2 \leq \frac{1}{2}\left(\sum_{k=1}^{n}|x_k - a|\right)^2.$$

6. P is a point inside of a given triangle ABC. D, E, F are the feet of the perpendiculars from P to the lines BC, CA, AB, respectively. Find all P for which $\frac{BC}{PD}+\frac{CA}{PE}+\frac{AB}{PF}$ is least (Problem from the 1981 International Mathematical Olympiad).

7. Let t be a root of the equation $x^2 - 3x + 1 = 0$.

   (a) For any given rational number a, find rational numbers b and c, such that $(t + a)(bt + c) = 1$;

   (b) Write $\frac{1}{t^2+2}$ in the form of $dt+e$, where d and e are rational numbers.

8. Suppose that integers $x_1, x_2, \ldots, x_n$ satisfy the following conditions:

   (a) $-1 \leq x_i \leq 2$, $i = 1, 2, \ldots, n$;
   (b) $x_1 + x_2 + \cdots + x_n = 19$;
   (c) $x_1^2 + x_2^2 + \cdots + x_n^2 = 99$.

   Find the maximum and minimum of $x_1^3 + x_2^3 + \cdots + x_n^3$.

9. Suppose that x and y are non-negative integers, such that $x + 2y$ is a multiple of 5, $x + y$ is a multiple of 3, and $2x + y \geq 99$. Find the minimum of $7x + 5y$.

10. Given are three positive real numbers a, b, c satisfying $abc + a + c = b$. Find the maximum value of the expression:

$$P = \frac{2}{a^2 + 1} - \frac{2}{b^2 + 1} + \frac{3}{c^2 + 1}.$$

(Problem of 1999 Vietnam Mathematical Olympiad).

Chapter 2

Proofs by Contradiction

An important method in mathematical proof is by contradiction. The basic ideal is this: you first assume that the proposition is false, and then use deductions to drive a conclusion contradicting the known facts.

Therefore, the procedure of proof by contradiction is

(a) Assumption of the contrary
(b) Reductions that lead to absurdity
(c) Conclusion.

Typically, the proof begins with the assumption that the proposition is false. It gives a contrary statement and uses deductions to show the contrary statement conflicts with axioms, given definitions, or conditions. There is no fixed pattern for the deduction steps, as the establishment of the contrary statement varies from problem to problem.

Here are some common ways to establish the contrary statement:

(1) By negation, such as A 'is not true', 'does not exist', 'is not equal to B', or 'cannot meet the given condition'.
(2) Use opposite relations on quantities, such as 'at most/at least', 'finite/infinite', or 'unique/several'.
(3) Use opposite terms in maths, such as 'coprime/have common divisor greater than 1', 'rational number/irrational number', 'there exists one A satisfying B/for any A, condition B cannot be satisfied', and so on.

We use the following examples to demonstrate the method of proof by contridiction.

Example 1. Nine mathematicians meet at an international conference. Among any three people, at least two of them share some common language. If none of them speaks more than three languages, show that there are at least three people sharing a common language.

Solution. Suppose, on the contrary, that each language is spoken by at most two persons. Therefore, in one language each person can only talk with at most one person.

Denote the mathematicians by $A_1, A_2, \ldots, A_9$. Since A_1 can speak at most three languages, by the assumption, he can speak to up to three people, and cannot speak to other 5 people, say A_2, A_3, A_4, A_5, and A_6. By the assumption again, A_2 can speak to at most three people, so he cannot talk with one of A_3, A_4, A_5, and A_6, say A_3. It follows that, A_1, A_2, and A_3 cannot communicate, a contradiction.

Example 2. Show that in an arbitrary triangle, there exist two side lengths a and b, satisfying

$$1 \leq \frac{a}{b} < \frac{1 + \sqrt{5}}{2}.$$

Solution. Suppose that the conclusion is false. Let $a \geq b \geq c$ be the side lengths of a triangle, and

$$\frac{a}{b} \geq \frac{1 + \sqrt{5}}{2}, \frac{b}{c} \geq \frac{1 + \sqrt{5}}{2}.$$

It follows that,

$$\frac{b + c}{a} = \frac{b}{a}\left(1 + \frac{c}{b}\right) \leq \frac{\sqrt{5} - 1}{2}\left(1 + \frac{\sqrt{5} - 1}{2}\right) = 1,$$

which contradicts the property $b + c > a$. The proof is complete.

Example 3. Given five segments. It is possible to buide a triangle of every subset of three of them. Prove that at least one of those is acute-angled (Problem from the 1970 All Soviet Union Mathematical Olympiad).

Solution. Let the lengths of five sticks be a, b, c, d, e, with $0 < a \leq b \leq c \leq d \leq e$. Suppose, on the contrary, any triangle formed by three sides of them is right-angled or obtuse. Then $d \geq c$, $c^2 \geq a^2 + b^2$, $a + b > e$, $e^2 \geq c^2 + d^2$.

Thus, $c^2 + d^2 \geq 2c^2 \geq 2(a^2 + b^2) \geq (a + b)^2 > e^2 \geq c^2 + d^2$, a contradiction.

Remark. In Example 3, very little information can be used to prove the result directly. Therefore, we use the method of contradiction to get more information that is useful.

Example 4. Show that among all six internal trisectrices of $\triangle ABC$, no three of them are concurrent.

Solution. Suppose three trisectrices intersect at point P. Then they must emanate from A, B, C, respectively. By the law of sine,

$$\frac{\sin \angle BAP}{\sin \angle ABP} \cdot \frac{\sin \angle CBP}{\sin \angle BCP} \cdot \frac{\sin \angle ACP}{\sin \angle CAP} = \frac{BP}{AP} \cdot \frac{CP}{BP} \cdot \frac{AP}{CP} = 1. \quad (1)$$

Let $A = 3\alpha$, $B = 3\beta$, $C = 3\gamma$. Clearly, $0 < \alpha, \beta, \gamma < \frac{\pi}{3}$, $\alpha + \beta + \gamma = \frac{\pi}{3}$.

By changing the internal angles if necessary, it only requires to consider two cases:

(a) $\angle BAP = \alpha$, $\angle CBP = \beta$, $\angle ACP = \gamma$.

Then, (1) becomes

$$\frac{\sin \alpha}{\sin 2\beta} \cdot \frac{\sin \beta}{\sin 2\gamma} \cdot \frac{\sin \gamma}{\sin 2\alpha} = 1,$$

which can be simplified as

$$\cos \alpha \cos \beta \cos \gamma = \frac{1}{8}.$$

This leads to contradiction as we have $\cos \alpha \cos \beta \cos \gamma > \left(\cos \frac{\pi}{3}\right)^3 = \frac{1}{8}$.

(b) $\angle BAP = \alpha$, $\angle CBP = \beta$, $\angle ACP = 2\gamma$.

Then, (1) becomes

$$\frac{\sin \alpha}{\sin 2\beta} \cdot \frac{\sin \beta}{\sin \gamma} \cdot \frac{\sin 2\gamma}{\sin 2\alpha} = 1.$$

Simplify the above equality to obtain

$$\cos \gamma = 2 \cos \alpha \cos \beta.$$

But

$$1 > \cos \gamma = 2 \cos \alpha \cos \beta = \cos(\alpha - \beta) + \cos(\alpha + \beta)$$

$$> 2 \cos(\alpha + \beta) > 2 \cos \frac{\pi}{3} = 1.$$

This is a contradiction.

Remark. In Example 4, the conclusion is given in the negative. We assume it to be true and derive a contradiction at the end. This is very common in the proof by contradiction.

Example 5. Is there a $\triangle ABC$, with side lengths $BC = a$, $CA = b$, $AB = c$, inscribed in a unit circle, and a number $p \in \mathbb{R}$, such that the roots of the equation $x^3 - 2ax^2 + bcx = p$ are exactly $\sin A$, $\sin B$, and $\sin C$?

Solution. The answer is negative. Suppose such a triangle and p exist. Then by Viète's theorem, we have

$$\begin{cases} 2a = \sin A + \sin B + \sin C, \\ bc = \sin A \sin B + \sin B \sin C + \sin C \sin A. \end{cases} \tag{1}$$

By the law of sine, we have

$$\frac{\sin A}{a} = \frac{\sin B}{b} = \frac{\sin C}{c} = \frac{1}{2}. \tag{2}$$

By (1) and (2), we see that

$$\begin{cases} 2a = (a + b + c)/2, \\ bc = (ab + bc + ca)/4. \end{cases}$$

They give

$$\begin{cases} 3a = b + c, \\ 3bc = a(b + c). \end{cases} \tag{3}$$

Thus, $3bc = a(b + c) = 3a^2$. So $a^2 = bc$. By (3) and $b < a + c$, we have $3a = b + c < a + 2c$, and hence $a < c$. Similarly, $a < b$. It follows that $a^2 < bc$, a contradiction.

Remark. To prove the non-existence of certain objects or quantities, we first assume that they exist, and try to use deduction to clarify that this is impossible. Then the desired conclusion is verified.

Example 6. We consider a prism which has the upper and inferior basis of pentagons: $A_1 A_2 A_3 A_4 A_5$ and $B_1 B_2 B_3 B_4 B_5$. Each of the sides of the two pentagons and the segments $A_i B_j$ with $i, j = 1, \ldots, 5$ is coloured either red or blue. In every triangle which has all sides coloured there exists one red side and one blue side. Prove that all the 10 sides of the two faces are coloured in the same colour (Problem from the 1979 International Mathematical Olympiad).

Solution. (a) Note that at least three sides of five sides $A_1 B_j$ ($j = 1, 2, 3, 4, 5$) have the same colour. In addition, the end points of the three sides on the bottom face must have two adjacent points. Without loss of generality, say $A_1 B_1$ and $A_1 B_2$ have the same colour, where B_1 and B_2 are adjacent.

(b) First, we show that all five sides of the top or bottom face have the same colour by contradiction. By symmetry of the prism, we need only to prove for the top face.

Suppose on the contrary, without loss of generality, $A_1 A_2$ and $A_1 A_5$ have different colours, $A_1 B_1$ and $A_1 B_2$ have the same colour as $A_1 A_2$, otherwise exchange A_2 with A_5. We see that the sides $A_2 B_1$, $A_2 B_2$, and $B_1 B_2$ of triangle $\triangle A_2 B_1 B_2$ have the same colour, since they are in triangles $\triangle A_1 A_2 B_1$, $\triangle A_1 A_2 B_2$, and $\triangle A_1 B_1 B_2$, respectively. A contradiction.

(c) Next, we show that the sides of top and bottom faces have the same colour. We also prove it by contradiction.

Without loss of generality, suppose that the colour of the sides of the top face is different from that of the bottom face. We have proved in (a) that the sides $A_1 B_1$ and $A_1 B_2$ have the same colour, but are different from the colour of $B_1 B_2$, that is $A_1 B_1$ and $A_1 B_2$ have the same colour as $A_1 A_2$. Then, we can derive the same contradiction as in (b).

Example 7. Let $a_0, a_1, \ldots$ be an infinite sequence of positive numbers. Show that the inequality $1 + a_n > 2^{1/n} a_{n-1}$ holds for infinitely many positive integers n (Problem from the 2001 International Mathematical Olympiad Shortlist).

Solution. Suppose, on the contrary, that the inequality $1 + a_n > 2^{1/n} a_{n-1}$ holds only for finitely many positive integers n. Let the largest such n be $M > 0$. Then, for any positive integer $n > M$,

$$1 + a_n \leq 2^{1/n} a_{n-1} \quad (n > M).$$

Equivalently,

$$a_n \leq 2^{1/n} a_{n-1} - 1 \quad (n > M). \tag{1}$$

By Bernoulli's inequality,

$$2^{1/n} = (1+1)^{1/n} \leq 1 + \frac{1}{n} = \frac{n+1}{n} \quad (n \geq 1). \tag{2}$$

It follows from (1) and (2) that

$$a_n \leq \frac{n+1}{n} a_{n-1} - 1 \quad (n > M).$$

That is

$$\frac{a_n}{n+1} - \frac{a_{n-1}}{n} \leq -\frac{1}{n+1} \quad (n > M). \tag{3}$$

Summing up (3) from $n = M+1$ to m, we obtain

$$\frac{a_m}{m+1} \leq \frac{a_M}{M+1} - \sum_{n=M+1}^{m} \frac{1}{n+1} \quad (m > M).$$

It is easy to see that $\lim\limits_{m \to +\infty} \sum\limits_{n=M+1}^{m} \frac{1}{n+1} = +\infty$. So, there exists $m_0 \geq M+1$, such that

$$\frac{a_{m_0}}{m_0+1} \leq \frac{a_M}{M+1} - \sum_{n=M+1}^{m_0} \frac{1}{n+1} < 0.$$

This is a contradiction.

Example 8. Given an acute-angled triangle ABC whose circumcentre is O. Let K be a point on BC, different from its midpoint. D is on the extension of AK. Lines BD and AC, CD and AB intersect at N, M, respectively. Prove that if $OK \perp MN$, then four points A, B, D, C are concyclic (Problem from the 2010 China High School Mathematical League).

Solution. We prove the statement by contradiction. If A, B, C, and D were not concyclic, let E be the intersect point of the circumcircle of $\triangle ABC$ and AD. The extension of BE beyond E and line AN meet at point Q. The extension of CE beyond E and line AM meet at point P. Connect points P and Q (Fig. 2.1).

Let r be the radius of the circumcircle of $\triangle ABC$. Then
$PK^2 = (PO^2 - r^2) + (KO^2 - r^2)$ (see Remark (a))

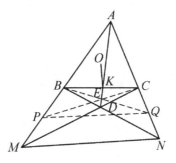

Figure 2.1

In addition, we have

$$QK^2 = (QO^2 - r^2) + (KO^2 - r^2).$$

Consequently,

$$PO^2 - PK^2 = QO^2 - QK^2.$$

Then, by vector representation,

$$(\vec{P} - \vec{O})^2 - (\vec{P} - \vec{K})^2 = (\vec{Q} - \vec{O})^2 - (\vec{Q} - \vec{K})^2.$$

This can be simplified to

$$0 = 2(\vec{O} - \vec{K})(\vec{P} - \vec{Q}).$$

That is, $OK \perp PQ$. Together with $OK \perp MN$ by the problem, we have $PQ \parallel MN$. Hence,

$$\frac{AQ}{QN} = \frac{AP}{PM}. \tag{1}$$

By Menelaus's theorem, we have

$$\frac{NB}{BD} \cdot \frac{DE}{EA} \cdot \frac{AQ}{QN} = 1, \tag{2}$$

and

$$\frac{MC}{CD} \cdot \frac{DE}{EA} \cdot \frac{AP}{PM} = 1. \tag{3}$$

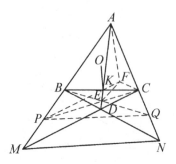

Figure 2.2

By (1), (2), and (3), we get

$$\frac{NB}{BD} = \frac{MC}{CD}.$$ (4)

Subtracting on both sides of (4) by 1, we have $\frac{ND}{BD} = \frac{MD}{CD}$, and thus $\triangle DMN \backsim \triangle DCB$. It follows that $\angle DMN = \angle DCB$, consequently, $BC // MN$. So, $OK \perp BC$. This implies that K is the midpoint of BC, a contradiction. Now the conclusion is verified.

Remark. (a) Extent PK to point F beyond K (see Fig. 2.2), such that

$$PK \cdot KF = AK \cdot KE.$$ (5)

Then four points P, E, F, and A are concyclic, thus, $\angle PFE = \angle PAE$. Moreover, $\angle PAE = \angle BCE$, since four points A, B, C, and E are concyclic. Thus, $\angle PFE = \angle BCE$. Consequently, four points E, C, F, and K are concyclic. It follows that

$$PK \cdot PF = PE \cdot PC.$$ (6)

Subtrating (5) from (6), we obtain

$$PK^2 = PE \cdot PC - AK \cdot KE = (QO^2 - r^2) + (KO^2 - r^2).$$

(b) The proof is similar when point E is on the extension of AD.

(c) This challenging problem has a deep projective geometry background. Moreover, a direct proof without using contradiction is possible. However, the converse statement is a known result, and it has a straightforward proof. For this reason, it is more natural for us to assume contrarily and prove the result by contradiction.

Example 9. Prove that equation $2x^3 + 5x - 2 = 0$ has exactly one real root r, and there exists a unique strictly increasing positive sequence $\{a_n\}$, such that $\sum_{n=1}^{+\infty} r^{a_n} = 2/5$ (Problem from the 2010 China High School Mathematical League).

Solution. Let $f(x) = 2x^3 + 5x - 2$. Then $f'(x) = 6x^2 + 5 > 0$, and $f(x)$ is a strictly increasing function. Since $f(0) = -2 < 0$, $f(1/2) = 3/4 > 0$. $f(x)$ has a unique real root r, and $r \in (0, 1/2)$. We can see that $\frac{2}{5} = \frac{r}{1-r^3} = \sum_{n=1}^{+\infty} r^{3n-2}$. Thus, the desired sequence is $a_n = 3n - 2$. $n = 1, 2, \ldots$. We prove the uniqueness of the sequence by contradiction. Suppose that there is another sequence $\{b_n\}$ of strictly increasing positive integers, and

$$\sum_{n=1}^{+\infty} r^{a_n} = \sum_{n=1}^{+\infty} r^{b_n}. \tag{1}$$

Eliminating all identical terms on both sides of (1), we get

$$\sum_{n=1}^{+\infty} r^{s_n} = \sum_{n=1}^{+\infty} r^{t_n}, \tag{2}$$

where $s_1 < s_2 < \cdots, t_1 < t_2 < \cdots$. Without loss of generality, suppose that $s_1 < t_1$.

Then $r^{s_1} < \sum_{n=1}^{+\infty} r^{s_n} = \sum_{n=1}^{+\infty} r^{t_n}$ and

$$1 < \sum_{n=1}^{+\infty} r^{t_n - s_1} \le \sum_{n=1}^{+\infty} r^n = \frac{1}{1-r} - 1 < \frac{1}{1-\frac{1}{2}} - 1 = 1.$$

This is a contradiction, and the conclusion follows.

Example 10. Let n be a positive integer and let $a_1, \ldots, a_k$ ($k \ge 2$) be distinct integers in the set $\{1, 2, \ldots, n\}$ such that n divides $a_i(a_{i+1} - 1)$ for $i = 1, \ldots, k - 1$. Prove that n does not divide $a_k(a_1 - 1)$ (Problem from the 2009 International mathematical Olympiad).

Solution. Since $n | a_i(a_{i+1} - 1)$, we have $a_i \equiv a_i a_{i+1} \pmod n$ for $i = 1, \ldots, k - 1$. Hence, $a_1 \equiv a_1 a_2 \equiv \cdots \equiv a_1 a_2 \cdots a_k \pmod n$.

Moreover, $a_1 a_2 \cdots a_k \equiv a_1 a_2 \cdots a_{k-2} a_k \equiv \cdots \equiv a_1 a_k \pmod n$.

Thus, $a_1 \equiv a_1 a_k \pmod n$.

Suppose, on the contrary, that $n | a_k(a_1 - 1)$. Then $a_1 a_k \equiv a_k \pmod n$. Which implies $a_1 \equiv a_k \pmod n$. As $1 \le a_1, \ldots, a_k \le n$ are distinct, this is impossible and the statement is verified.

Exercises

1. Prove that for any real numbers x, y, and z the following three inequalities cannot hold simultaneously

$$|x| < |y - z|, \quad |y| < |z - x|, \text{ and } |z| < |x - y|.$$

2. Four points E, F, G, H are on the four sides of a unit square $ABCD$, respectively. Prove that at least one side of the quadrilateral $EFGH$ has length $\geq \sqrt{2}/2$.

3. (1) Are there positive integers m and n, such that $m(m+2) = n(n+1)$?
   (2) Let $k(\geq 3)$ be a given positive integer. Are there positive integers m and n, such that $m(m + k) = n(n + 1)$?

4. Let a, b, c be real numbers with $a > 2000$. Show that the inequation $|ax^2 + bx + c| \leq 1000$ has at most two integer solutions.

5. Is there a triangle with sides of integer lengths such that the length of the shortest side is 2007 and that the largest angle is twice the smallest? (Problem from the 2007 Western China Mathematical Olympiad).

6. Let A be the set of reciprocals of all positive integers. Are there infinitely many numbers $a_1, a_2, \ldots$ in A, such that $\frac{a_i}{i} + \frac{a_j}{j} \in A$, for all $i, j \in \mathbb{N}^+$? Justify your conclusion.

7. A certain organization has n members, and it has $n + 1$ three-member committees, no two of which have identical membership. Prove that there are two committees which share exactly one member (Problem from the 1979 USA Mathematical Olympiad).

8. Let $a_1, a_2, \ldots$ be any permutation of all positive integers. Prove that there exist infinitely many positive integers i, such that $\gcd(a_i, a_{i+1}) \leq 3i/4$ (Problem from the 2011 China Team Selection Test).

Chapter 3

Inductive Methods

Suppose that for every integer $n \geq 1$ there is a statement $P(n)$. Such statements can often be proved by the method of mathematical induction. Since this method can also be applied to fields other than mathematics, we call it induction instead of mathematical induction.

(a) The basic form of induction (the first form of induction): if we can prove the following two properties:

   (I) The statement $P(1)$ is true.

   (II) For each integer $n \geq 1$, if $P(n)$ is true, then $P(n+1)$ is true.
   Then for all integers $n \geq 1$, the statement $P(n)$ is true.

(b) The second form of induction: if we can prove the following two properties:

   (I) The statement $P(1)$ is true.

   (II) For each integer $n \geq 1$, if $P(k)$ is true for every integer k with $1 \leq k \leq n$, then $P(n+1)$ is true.
   Then for all integers $n \geq 1$, the statement $P(n)$ is true.

(c) The third form of induction: if we can prove the following two properties:

   (I) The statements $P(n_0)$, $P(n_0 + 1), \ldots, P(n_0 + k - 1)$ are true.

   (II) For each integer $n \geq n_0$, if $P(n)$ is true, then $P(n+k)$ is true.
   Then for all integers $n \geq n_0$, the statement $P(n)$ is true. (Integer n_0 is called the induction basis and integer $k \geq 1$ is called the induction span).

(d) Reverse induction. It was first used by the famous French mathematician A. Cauchy. He proved by reverse induction that the arithmetic average of n positive real numbers is greater than or equal to the geometric mean of these n numbers.

If we can prove the following two properties:

(I) $P(n)$ is true for infinitely many integers $n \geq 1$.
(II) For each integer $n \geq 1$, if $P(n + 1)$ is true, then $P(n)$ is true.
   Then for all integers $n \geq 1$, the statement $P(n)$ is true.
   Usually, we say that (I) is the base of induction, and $n = 1$ is the induction basis; say that (II) is the inductive step of induction. None of them can be missing.

Remark. Obviously, in (a), (b), and (d) forms of induction, we could replace induction basis 1 by any integer n_0.

(e) Induction on two parameters. Suppose that for every integer $n \geq 1$ and every integer $m \geq 1$ we are given a statement $P(n, m)$. If we can prove the following two properties:

(I) For any integer $m \geq 1$, $P(1, m)$ is true, and for any integer $n \geq 1$, $P(n, 1)$ is true.
(II) For each integer $n \geq 1$ and $m \geq 1$, if $P(n + 1, m)$ and $P(n, m + 1)$ are true, then $P(n + 1, m + 1)$ is true.
   Then for all integers $n \geq 1$ and $m \geq 1$, the statement $P(n, m)$ is true.

The method of induction can be used widely. And in many cases it has some variants and involves many techniques, such as proposition enhancement, basis and span selection, and so on.

Example 1. Let $\{x_k\}$ be a sequence of real numbers satisfying

$$\sum_{k=0}^{n} x_k^3 = \left(\sum_{k=0}^{n} x_k \right)^2 \quad \text{for any } n \in \mathbb{N}.$$

Prove that for any given $n \in \mathbb{N}$, there exists an integer m such that

$$\sum_{k=0}^{n} x_k = \frac{m(m + 1)}{2}.$$

Solution. We prove by induction. If $n = 0$, then by $x_0^3 = x_0^2$, we have $x_0 = 0$, or 1. Then we can take $m = 0$ or 1, respectively. Suppose that for $n \geq 0$, $\sum_{k=0}^{n} x_k^3 = \left(\sum_{k=0}^{n} x_k \right)^2$, there exists an integer m such that

$\sum_{k=0}^{n} x_k = \frac{m(m+1)}{2} =: c.$ Then $\sum_{k=0}^{n} x_k^3 = c^2.$ Thus,

$$c^2 + x_{n+1}^3 = \sum_{k=0}^{n+1} x_k^3 = \left(\sum_{k=0}^{n+1} x_k\right)^2 = (c + x_{n+1})^2.$$

That is $x_{n+1}(x_{n+1}^2 - x_{n+1} - m(m+1)) = 0$, so $x_{n+1} = 0, -m,$ or $m+1$. Therefore,

$$\sum_{k=0}^{n+1} x_k = \begin{cases} \dfrac{m(m+1)}{2}, & \text{if } x_{n+1} = 0, \\[2mm] \dfrac{(m-1)m}{2}, & \text{if } x_{n+1} = -m, \\[2mm] \dfrac{(m+1)(m+2)}{2}, & \text{if } x_{n+1} = m+1. \end{cases}$$

This completes the proof.

Remark. In the process of induction, the difficulty and the key lie on the inductive step, in which the relation between $P(n)$ and $P(n+1)$ should be exposed. According to the problem, we either start from $P(n)$ and transit to $P(n+1)$, or separate $P(n)$ from $P(n+1)$. Sometimes, transformations to the proposition or expressions are needed repeatedly. After all, we should use the inductive assumption thoroughly.

Example 2. Let 2^n balls be divided into several piles. We call it a *move*: for any two piles A and B of balls, if A has p ball(s) and B has q ball(s), and $p \geq q$, then move q ball(s) from A to B. Prove that we can have all 2^n balls in one pile by a finite number of *moves*.

Solution. We prove by induction on n.

If $n = 1$, there are only two balls, so we can complete the task by one *move*.

Suppose that for $n \geq 1$, the conclusion is true. For 2^{n+1} balls, the number of piles with an odd number of ball(s) is even. So, we make them in pairs, and move balls between each pair to make all numbers even. If we bind every 2 balls together as one *big ball*, then there are all together 2^n *big balls*. So, this completes the proof.

Example 3. Prove that the nth prime $p_n \leq 2^{2^n - n}$, and the $<$ holds for $n \geq 2$.

Solution. If $n = 1$, $p_1 = 2 = 2^1$. If $n = 2$, $p_2 = 3 < 2^{4-2} = 4$. So $p_1 p_2 < 2^{2-1+2^2-2}$. Suppose that $p_k < 2^{2^k-k}$ for $2 \leq k \leq n$, then

$$\prod_{k=1}^{n} p_k < 2^{-n(n+1)/2 + \sum_{k=1}^{n} 2^k} = 2^{2^{n+1}-1-n(n+1)/2}, \text{ for } n \geq 2.$$

Thus,

$$p_{n+1} \leq 1 + \prod_{k=1}^{n} p_k \leq 2^{2^{n+1}-1-n(n+1)/2} < 2^{2^{n+1}-(n+1)}, \text{ for } n \geq 2.$$

This completes the proof.

Remark. In Example 3, we use the second form of induction. So, the more the assumptions, the easier it is to get the result.

Example 4. Let $0 < a < 1$, $x_0 = 1$, $x_{n+1} = \frac{1}{x_n} + a$, $n \in \mathbb{N}$. Show that $x_n > 1$ for all $n > 0$.

Solution 1. If $n = 1$, $x_1 = 1 + a > 1$. If $n = 2$, $x_2 = \frac{1}{1+a} + a = 1 + \frac{a^2}{1+a} > 1$. Suppose that for $n \geq 1$, $x_n > 1$, then $0 < x_{n+1} = \frac{1}{x_n} + a < 1 + a$. Hence,

$$x_{n+2} = \frac{1}{x_{n+1}} + a > \frac{1}{1+a} + a = 1 + \frac{a^2}{1+a} > 1.$$

This completes the induction.

Solution 2. We prove a stronger conclusion: $1 < x_n < 1 + a$, for all $n > 1$.
    For $n = 2$, $x_2 = \frac{1}{1+a} + a = \frac{1+a+a^2}{1+a} = 1 + \frac{a^2}{1+a}$, so $1 < x_2 < 1 + a$. Suppose that for $n \geq 2$, we have $1 < x_n < 1 + a$, and then $1 < \frac{1+a+a^2}{1+a} < \frac{1}{x_n} + a < 1 + a$.
    That is $1 < x_{n+1} < 1 + a$. This completes the induction.

Remark. The above problem cannot be proved by the first form of induction. So, in Solution 1, we use the third form of induction with inductive step 2. But in this case, we should check in the base step for $n = 1$ and $n = 2$. In Solution 2, by strengthening the conclusion, we can use the first form of induction.

Example 5. Let $a_n = \sum_{k=1}^{n} \frac{1}{k}$. Prove that

$$a_n^2 > 2 \sum_{k=2}^{n} \frac{a_k}{k}, \text{ for } n \geq 2.$$

Solution. Strengthen the conclusion to

$$a_n^2 > \frac{1}{n} + 2\sum_{k=2}^{n} \frac{a_k}{k}, \text{ for } n \geq 2.\tag{1}$$

If $n = 2$, $a_2^2 = \left(1 + \frac{1}{2}\right)^2 = \frac{9}{4} > \frac{1}{2} + 2\left(\frac{1}{2}\left(1 + \frac{1}{2}\right)\right) = 2$.
Suppose that (1) is true for $n \geq 2$, then

$$a_{n+1}^2 = \left(a_n + \frac{1}{n+1}\right)^2 = \frac{2a_n}{n+1} + \frac{1}{(n+1)^2} + a_n^2$$

$$> \frac{2}{n+1}\left(a_{n+1} - \frac{1}{n+1}\right) + \frac{1}{(n+1)^2} + \frac{1}{n} + 2\sum_{k=2}^{n} \frac{a_k}{k}$$

$$= \frac{1}{n} - \frac{1}{(n+1)^2} + 2\sum_{k=2}^{n+1} \frac{a_k}{k} > \frac{1}{n+1} + 2\sum_{k=2}^{n+1} \frac{a_k}{k}.$$

This completes the induction.

Example 6. There are 2005 young people sitting around a (large!) round table. Of these, at most 668 are boys. We say that a girl G is in a *strong position*, if, counting from G to either direction at any length, the number of girls is always strictly larger than the number of boys (G herself is included in the count). Prove that in any arrangement, there is always a girl in a *strong position* (Problem from the 2005 Nordic Mathematical Contest).

Solution. We strengthen the proposition to a more general problem:

Suppose n boys and $2n + 1$ girls are sitting around a large round table. Then there is a girl, counting from her in the anticlockwise direction at any length, the number of girls is always larger than the number of boys.

Label the positions of the seats clockwise from 1 to $3n + 1$.

If $n = 1$, without loss of generality, we may let the boy be 4, girls be 1, ·2, and 3. Then, the girl 3 is in the *strong position*.

Suppose that the proposition is true for $n \geq 1$. Let us consider $n + 1$ boys. Since there are more girls than boys, there must be two girls who are adjacent and neighbouring with a boy on the right. Without loss of generality, let the boy position be 1; then, 2 and 3 are girls. Ignore these three persons, then, with n boys and $2n + 1$ girls left, by the inductive assumption, there is a girl A at a *strong position* m. We can see position m is also a *strong position* when positions 1, 2, and 3 are included. This completes the proof of the general problem.

In case of counting in the other direction, the proof is similar.

Return to the original problem, and take $n = 668$. If the number of boys is less than 668, the conclusion is also true, since we can regard some girls as boys.

Remark. This is a typical problem solved by being strengthened to a general one.

At the inductive step, the stronger the proposition, and the more the conditions in case n, the stronger the proof becomes.

Sometimes we reinforce or generalize the conclusion of a proposition. It seems that the proposition is more difficult to deal with because the conclusion is strengthened. However, on the contrary, sometimes it is easier to solve than the original case. This technique is often used in induction.

Soviet mathematician Aleksandr Yakovlevich Khinchin once said: 'in the proof of mathematical induction, the stronger the proposition, the more the conditions needed at assumption step, however, the more are to be proved at the induction step. But for many problems, the greater number of conditions are more important in the proof.'

Example 7. Prove that the Diophantine problem $x^2 + y^2 = z^n$ has solutions of positive integers for any positive integer n.

Solution. For $n = 1$, $(x, y, z) = (1, 1, 2)$ is a solution. For $n = 2$, $(x, y, z) = (3, 4, 5)$ is a solution. Suppose that $(x, y, z) = (x_0, y_0, z_0)$ is a solution for $n \geq 1$; then $(x, y, z) = (z_0 x_0, z_0 y_0, z_0)$ is a solution for $n + 2$.

Example 8. For what positive integer n is the set $\{1, 2, \ldots, n\}$ the union of 5 mutually disjoint subsets with the same sum of elements?

Solution. Since $1 + 2 + \cdots + n = n(n+1)/2$, and $5 | n(n + 1)/2$, we see that $n = 5k$ or $5k - 1$, $k \in \mathbb{N}^+$.

Obviously, $k = 1$ is not an answer. In the following, we shall prove by induction that either $n = 5k$ or $n = 5k - 1$, $k \geq 2$ satisfy the conditions.

If $k = 2$, $n = 5k - 1 = 9$, the subsets are $\{9\}$ and $\{j, 9 - j\}$, $j = 1, \ldots, 4$.

If $k = 2$, $n = 5k = 10$, the subsets are $\{j, 11 - j\}$, $j = 1, \ldots, 5$.

If $k = 3$, $n = 5k - 1 = 14$, the subsets are $\{1, 2, 3, 4, 5, 6\}$ and $\{j, 21 - j\}$, $j = 7, \ldots, 10$.

If $k = 3$, $n = 5k = 15$, the subsets are $\{1, 2, 3, 5, 6, 7\}$, $\{4, 8, 12\}$, and $\{j, 24 - j\}$, $j = 9, 10, 11$.

Suppose that for $k \geq 2$, $n = 5k$ or $n = 5k - 1$ meets the requirements, that is $\{1, \ldots, n\} = \bigcup_{j=1}^{5} A_j$, $A_i \cap A_j = \emptyset, i \neq j$, $i, j = 1, \ldots, 5$ and the sums of the elements of A_j are equal. Then let $B_j = A_j \cup \{n+j, n+11-j\}$, $j = 1, \ldots, 5$. Hence, for $k+2$, $\{1, \ldots, n, n+1, \ldots, n+10\} = \bigcup_{j=1}^{5} B_j$ meets the requirements.

Remark. In the above problem, we take 2 and 3 as the bases. Sometimes, more bases are needed.

Example 9. Let f be a function defined on $[0, +\infty)$ with $f(0) = 0$ and satisfying $|f(x) - f(y)| \leq (x - y)f(x)$, for any $x \geq y \geq 0$. Find f.

Solution. We prove by induction that for any positive integer n, $f(x) = 0$, for $x \in \left[\frac{n-1}{2}, \frac{n}{2}\right)$. Therefore, $f(x) \equiv 0$, for all $x \geq 0$.

If $n = 1$, take $y = 0$, $x \in [0, 1/2)$, and then $|f(x)| \leq xf(x) \leq x|f(x)| < |f(x)|/2$. That is $f(x) = 0$, for $x \in [0, 1/2)$.

Suppose that for $n \geq 1$, $f(x) = 0$, for $x \in \left[\frac{n-1}{2}, \frac{n}{2}\right)$. Then, for $x \in \left[\frac{n}{2}, \frac{n+1}{2}\right)$, take $y = x - 1/2 \in \left[\frac{n-1}{2}, \frac{n}{2}\right)$, $f(y) = 0$, and $|f(x)| \leq f(x)/2 \leq |f(x)|/2$. That is, $f(x) = 0$, for $x \in \left[\frac{n}{2}, \frac{n+1}{2}\right)$. This completes the induction.

Remark. The above proof is an example of the application of induction to real numbers which are not countable. That is, there does not exist a sequence that goes through all positive real numbers. So, we take the intervals $[(n-1)/2, n/2)$ instead, so that we can use the inductive method.

Example 10. Let real valued function $f(n, m)$ satisfy $f(n, 1) = 1 = f(1, m)$ for $m, n \in \mathbb{N}^+$, and $f(n, m) \leq f(n, m-1) + f(n-1, m)$, for $m, n \geq 2$. Prove the proposition $P(n, m) : f(n, m) \leq \dbinom{m+n-2}{n-1}$.

Solution. We prove by induction on two parameters $m, n \in \mathbb{N}^+$.
$$f(n, 1) = 1 = \binom{1+n-2}{n-1}, \quad f(1, m) = 1 = \binom{m+1-2}{1-1}, \text{ that is}$$
$P(n, 1)$ and $P(1, m)$ are true.

Suppose that for $m, n \geq 1$, $P(n+1, m)$ and $P(n, m+1)$ are true. Then,
$$f(n+1, m+1) \leq f(n+1, m) + f(n, m+1) \leq \binom{m+n-1}{n} + \binom{m+n-1}{n-1} = \binom{m+n}{n}.$$
That is $P(n+1, m+1)$ is true. This completes the induction.

Remark. There is a direct proof for Example 10. Let $F(n, m) = \binom{m+n-2}{n-1}$. We see that $F(n, 1) = f(n, 1) = 1$, $F(1, m) = f(1, m) = 1$ for $m, n \in \mathbb{N}^+$, and by the combinatorial property, $F(n, m) = F(n, m-1) + F(n-1, m)$, for $m, n \geq 2$.

Since $f(n, m) \leq f(n, m-1) + f(n-1, m)$ for $m, n \geq 2$, $f(m, n) \leq F(m, n)$.

Example 11. Let k be a positive integer. Prove that one can partition the set $\{0, 1, 2, \ldots, 2^{k+1} - 1\}$ into two disjoint subsets $\{x_1, x_2, \ldots, x_{2^k}\}$ and $\{y_1, y_2, \ldots, y_{2^k}\}$, such that $\sum_{i=1}^{2^k} x_i^m = \sum_{i=1}^{2^k} y_i^m$ for all $m \in \{1, 2, \ldots, k\}$. (Problem from the 2005 China Team Selection Test).

Solution. We prove by induction on k. If $k = 1$, let $\{x_1, x_2\} = \{0, 3\}$, $\{y_1, y_2\} = \{1, 2\}$.

Suppose that the proposition is true for $k \geq 1$. That is, the set $\{0, 1, 2, \ldots, 2^{k+1} - 1\}$ can be the union of two disjoint subsets $\{x_1, x_2, \ldots, x_{2^k}\}$ and $\{y_1, y_2, \ldots, y_{2^k}\}$, such that

$$\sum_{i=1}^{2^k} x_i^m = \sum_{i=1}^{2^k} y_i^m \text{ for any integer } 1 \leq m \leq k. \tag{1}$$

Now, for $k+1$, let

$$A_{k+1} = \{x_1, x_2, \ldots, x_{2^k}, 2^{k+1} + y_1, 2^{k+1} + y_2, \ldots, 2^{k+1} + y_{2^k}\},$$

$$B_{k+1} = \{y_1, y_2, \ldots, y_{2^k}, 2^{k+1} + x_1, 2^{k+1} + x_2, \ldots, 2^{k+1} + x_{2^k}\}.$$

Then $A_{k+1} \cup B_{k+1} = \{0, 1, \ldots, 2^{k+2} - 1\}$ and $A_{k+1} \cap B_{k+1} = \emptyset$. Now we have to verify that

$$\sum_{i=1}^{2^k} x_i^m + \sum_{i=1}^{2^k} (2^{k+1} + y_i)^m = \sum_{i=1}^{2^k} y_i^m + \sum_{i=1}^{2^k} (2^{k+1} + x_i)^m, \text{ for } 1 \leq m \leq k+1. \tag{2}$$

Expand (2) and eliminate the identical terms on both sides, (2) is equivalent to

$$\sum_{j=0}^{m-1} \binom{m}{j} (2^{k+1})^{m-j} \left(\sum_{i=1}^{2^k} x_i^j - \sum_{i=1}^{2^k} y_i^j \right) = 0, \text{ for } 1 \leq m \leq k+1. \tag{3}$$

Since $\sum_{i=1}^{2^k} x_i^j - \sum_{i=1}^{2^k} y_i^j = 0$ for $j = 1, 2, \ldots, k$ by (1), and

$$\sum_{i=1}^{2^k} x_i^0 - \sum_{i=1}^{2^k} y_i^0 = 2^k - 2^k = 0,$$

(3) holds.

Remark. In this problem, the inductive step is a constructive proof. It seems hard to have a non-constructive proof.

Conclusion. We use dominoes to illustrate the principle of induction.

For the first form of induction, the basis of induction is just as the fall of the first domino, and the inductive step is just as the falling of the next domino after the previous one. Every domino falls when the induction is completed.

For the second form of induction, first let the first domino fall, then the falling of the first n dominoes makes the $(n + 1)$ domino fall. Eventually all the dominoes will fall.

For the reverse induction, first make sure that each domino's falling can cause the previous ones to fall, next we should show that infinitely many dominoes are falling.

Exercises

1. For any positive integer $n \in \mathbb{N}^+$, $a_n > 0$, and $\sum_{i=1}^n a_i^3 = \left(\sum_{i=1}^n a_i\right)^2$, prove that $a_n = n$ (Problem from the 1989 China High School Mathematical League).

2. Define sequence $\{a_n\}$ by $a_{n+1} = 3^{a_n}$, $n \in \mathbb{N}^+$, with $a_1 = 3$. And define sequence $\{b_n\}$ by $b_{n+1} = 8^{b_n}$, $n \in \mathbb{N}^+$, with $b_1 = 8$. Prove that $a_{n+1} > b_n$ for all $n \in \mathbb{N}^+$.

3. Show that we can cut a square into n (≥ 6) smaller squares.

4. Let $M = \sum_{j=1}^s 2^{a_j}$, where a_j, $j = 1, 2, \ldots, s$ are distinct positive integers. Prove that $\sum_{j=1}^s 2^{a_j/2} < (1 + \sqrt{2})\sqrt{M}$ (Problem from the 2009 Jiangsu Province Preliminary for China High School Mathematical League).

5. Let $\{a_n\}$ be an increasing sequence of positive integers with $a_2 = 2$, and $a_{mn} = a_m a_n$ for any $m, n \in \mathbb{N}^+$. Show that $a_n = n$.

6. Prove that the equation $x_1^3 + x_2^3 + \cdots + x_n^3 = y^3$ has a positive solution $(x_1, x_2, \ldots x_n, y)$ with $x_1 < x_2 < \cdots < x_n$ for any given integer $n \geq 3$.

7. Prove that we can find an infinite set of positive integers of the form $2^n - 3$ (where $n > 1$ is a positive integer) every pair of which is relatively prime (Problem from the 1971 International Mathematical Olympiad).

8. Let $n > 0$ be an integer. We are given a balance and n weights of weight $1, 2, 4, \ldots, 2^{n-1}$. We are to place each of the n weights on the balance, one after another, in such a way that the right pan is never

heavier than the left pan. At each step, we choose one of the weights
that has not yet been placed on the balance, and place it on either the
left pan or the right pan, until all of the weights have been placed.

Determine the number of ways in which this can be done (Problem
from the 2011 International Mathematical Olympiad).

9. A set A of positive integers satisfies the following conditions:

   (1) A has 3 or more elements;
   (2) If $a \in A$, then all factors of a are in A;
   (3) If $a \in A$, $b \in A$, $1 < a < b$, then $1 + ab \in A$.

   Prove that $A = \mathbb{N}^+$.

10. Prove that, any n points in $\triangle ABC$ with $\angle C = 90°$ can be properly
    labeled as $P_1, \ldots, P_n$, such that $(P_1 P_2)^2 + (P_2 P_3)^2 + \cdots + (P_{n-1} P_n)^2 \leq (AB)^2$.

11. In a boarding school, there are 512 students and 256 rooms (2 people in
    each room). For every student there is a set of 9 subjects the student
    is interested in. Any two students have different sets of subjects.

    Prove that all students can queue up in a circle in such a way that
    every pair of roommates is neighbouring in the circle, and for any two
    neighbouring students in the circle who are not roommates there are
    exactly eight common subjects. (Problem from the 2010 All-Russian
    Mathematical Olympiad).

Chapter 4

The Drawer Principle

Drawer principle (also called the pigeonhole principle or the Dirichlet principle), firstly proposed by German mathematician Dirichlet, is a fundamental principle in combinatorial mathematics,

Put 10 apples into 9 drawers, then there is a drawer that has at least 2 apples. Put 10 apples into 3 drawers, then there is a drawer that has at least 4 apples. Put 9 apples into 3 drawers, then there is a drawer that has at least 3 apples. The reasons seem simple, but this is a very important method in solving problems of existence. The usual forms of the drawer principle are as follows:

Drawer principle 1. If we put $n + 1$ things into n drawers, then there is a drawer with at least 2 things.

Drawer principle 2. If we put m things into n drawers, then there is a drawer with at least $\lceil \frac{m}{n} \rceil = \lfloor \frac{m-1}{n} \rfloor + 1$ things. And there is also a drawer with at most $\lfloor \frac{m}{n} \rfloor$ things. Here and hereafter $\lfloor x \rfloor$ means the greatest integer not greater than x and $\lceil x \rceil$ means the least integer not less than x.

Drawer principle 3. If we put infinitely many things into finitely many drawers, then there is at least one drawer that has infinitely many things.

In fact, drawer principle 1 is a special case of drawer principle 2.

The key to successfully solving problems by using the drawer principle lies in how to design the *drawer*.

When using the drawer principle, sometimes we use the equivalent 'average principle', and 'overlapping principle' for geometric problems.

We always use the drawer principle for existence problems. Example 1 consists of problems ranging from easy to hard.

Example 1. For set $S = \{1, 2, \ldots, 100\}$, try to prove the following:

(1) If $A \subset S$ has 51 members, then there exist $i, j \in A$, such that $i - j = 50$.
(2) If $B \subset S$ has n members ($2 \leq n \leq 100$), then there exist $i, j \in B$ such that $0 < i - j < \frac{100}{n-1}$.
(3) If $C \subset S$ has 51 members, then there exist $i, j \in C$, $i < j$, such that $i | j$.
(4) If $D \subset S$ has 75 members, then there exist four members of D that are mutually coprime.

Solution. (1) Let $A_k = \{k, k + 50\}$, $k = 1, 2, \ldots, 50$. Since $S = \bigcup_{k=1}^{50} A_k$, by the drawer principle, there are two members of A in some A_k, then $i = k + 50$, $j = k$, thus, $i - j = 50$.

(2) Denote $\alpha = \frac{99}{n-1}$, and let $B_k = \{x \in S | 1 + (k - 1)\alpha \leq x \leq 1 + k\alpha\}$, $k = 1, 2, \ldots, n - 1$. Since $S = \bigcup_{k=1}^{n-1} B_k$, by the drawer principle, there is a subset B_k that has two members i, j in B. So $0 < i - j \leq \alpha < \frac{100}{n-1}$.

(3) Let $C_k = \{x \in S | x = 2^p(2k - 1), p \in \mathbb{N}\}$, $k = 1, 2, \ldots, 50$. Since $S = \bigcup_{k=1}^{50} C_k$, by the drawer principle, there is a C_k that has members i, j, $i < j$ (in 51 members of) C. So, $i | j$.

(4) First, we construct four subsets D_1, D_2, D_3, and D_4 of S, each with mutually coprime numbers. D_1 is the set of integers 1 and all 25 primes less than 100,

$$D_1 = \{1, 2, 3, 5, 7, 11, \ldots, 89, 97\}, \ D_2 = \{2 \times 47, 3 \times 31, 5 \times 19, 7 \times 13\},$$

$$D_3 = \{2 \times 43, 3 \times 29, 5 \times 17, 7 \times 11\}, \ D_4 = \{2 \times 41, 3 \times 23, 5 \times 13, 7^2\}.$$

Since $T = \bigcup_{i=1}^{4} D_i$ has 38 members, $S \backslash T$ has 62 members. We see that there are at least 13 members of the 75 members of D are in T. By the drawer principle, there is a set D_i that has at least four members.

Example 2. Put $1, 2, \ldots, 10$ on a circle in any order.

(1) Prove that there are three adjacent numbers with sum not less than 18.
(2) Prove that there are three adjacent numbers with sum not greater than 15.

Solution. (1) Suppose that the order of the numbers is $1, a_1, a_2, \ldots, a_9$ as shown in Fig. 4.1. Since the three sums of $a_1 + a_2 + a_3, a_4 + a_5 + a_6$, $a_7 + a_8 + a_9$ add up to 54, by the drawer principle, there is at least one sum that is not less than $54/3 = 18$.

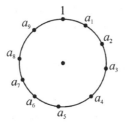

Figure 4.1

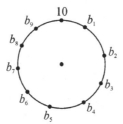

Figure 4.2

(2) Suppose that the order of the numbers is $10, b_1, b_2, \ldots, b_9$ as shown in Fig. 4.2. Since the sum of three sums of $b_1 + b_2 + b_3, b_4 + b_5 + b_6$, $b_7 + b_8 + b_9$ is 45, by the drawer principle, there is at least one sum not greater than $45/3 = 15$.

Remark. The proof above uses the average principle.

Example 3. Let m be a given positive integer. Let A_i $(i = 1, 2, \ldots, m)$ be sets of integers such that $\bigcup_{i=1}^{m} A_i = \mathbb{N}^+$, prove that there is a set A_i, $(1 \le i \le m)$ that contains multiples of any positive integer.

Solution. Consider the sequence $\{n!\}$. Since $n! \in \mathbb{N}^+ = \bigcup_{i=1}^{m} A_i$, by the drawer principle, there is at least one set A_i, $(1 \le i \le m)$ containing infinitely many terms of $\{n!\}$. Thus, for any integer n, there exists an integer $n_1 \ge n$, such that $n_1! \in A_i$, where $n_1!$ is a multiple of n.

Remark. In the above proof, the drawer principle 3 is used.

Example 4. How many numbers can be selected from integers 1 to 16, such that any three of them are not mutually coprime?

Solution. Taking multiples of 2 and 3 from 1 to 16, there are 11 numbers: $2, 3, 4, 6, 8, 9, 10, 12, 14, 15$, and 16.

So, any three of them are not mutually coprime.

Furthermore, we shall prove that for any 12 numbers in 1 to 16, there must be three numbers that are mutually coprime.

In fact, take a coprime integer set $A = \{1, 2, 3, 5, 7, 11, 13\}$. By the drawer principle, taking 12 numbers from 1 to 16, at least three numbers are selected from A.

Therefore, the answer is 11.

Remark. This problem is similar to Example 1(4), but we must explain that 11 is the upper bound. As for Example 1(4), we can find a subset of $S = \{1, 2, \ldots, 100\}$ with 74 members that does not satisfy the conditions of the problem, so 75 is the minimum that is desired by the problem.

Example 5. Let T be the set of all positive factors of 60^{100}. Let $S \subset T$, such that any number in S is not a multiple of any other number in S. Find the maximum of the number of members of S.

Solution. Since $60 = 2^2 \cdot 3 \cdot 5$, we have $T = \{2^a 3^b 5^c | 0 \le a \le 200, 0 \le b, c \le 100\}$. Let $S = \{2^{200-b-c} 3^b 5^c | 0 \le b, c \le 100\}$. It is easy to see that $S \subset T$ and S has 101^2 members.

We now show that S satisfies the requirement of the problem.

First, we show that S has the maximal number of members.

Since there are 101^2 different pairs (b, c), any subset of T with more than 101^2 members must contain two members of the form $2^{a_1} 3^b 5^c$ and $2^{a_2} 3^b 5^c$, where $a_1 \ne a_2$. So, one number is a multiple of the other.

Next, we show that in set S, any member is not a multiple of any other member.

Let $2^{200-b-c} 3^b 5^c$ be a multiple of $2^{200-i-j} 3^i 5^j$, then $b \ge i, c \ge j$ and $200 - b - c \ge 200 - i - j$. Therefore, $b = i$, $c = j$, and they refer to the same number.

Remark. The technique of making 'drawers' above is similar to Example 1(3).

Example 6. Suppose that $0 < a_1 < a_2 < \cdots < a_{2011}$. Prove that there exists i, such that

$$a_{i+1} - a_i < \frac{(1 + a_i)(1 + a_{i+1})}{2010}.$$

Solution. Put 2011 numbers $x_i = \frac{2010}{1+a_i}$, $i = 1, 2, \ldots, 2011$ (we see that $2010 > x_1 > x_2 > \cdots > x_{2011} > 0$) into 2010 intervals: $(0, 1], (1, 2], (2, 3], \ldots, (2008, 2009], (2009, 2010)$. By the drawer principle, there is an interval containing two numbers, say x_i, x_{i+1}, so, $0 < x_i - x_{i+1} < 1$, that is

$$a_{i+1} - a_i < \frac{(1 + a_i)(1 + a_{i+1})}{2010}.$$

Remark. The technique of making *drawers* above is similar to Example 1(2), but one should make a transformation first.

Example 7. A chess master will warm up for 77 days for a competition. He plans to play at least one game every day but at most 12 games a week. Prove that there is a positive integer n, such that he will play 21 games in n consecutive days.

Solution. Let a_i be the total number of games played by the master in the first i days, $i = 1, 2, \ldots, 77$. Since he will play at least one game each day, $1 \leq a_1 < a_2 < \cdots < a_{77}$.

And since he will play at most 12 games each week, $a_{77} \leq 12 \times 77/7 = 132$. Hence,

$$22 \leq a_1 + 21 < a_2 + 21 < \cdots < a_{77} + 21 \leq 153.$$

Consider the sequence of 154 terms $a_1, a_2, \ldots, a_{77}, a_1 + 21, a_2 + 21, \ldots, a_{77} + 21$. By the drawer principle, there exist i, j, such that $a_i = a_j + 21$. Therefore, the master will play 21 games in $n = i - j$ consecutive days from the $(j + 1)$th day to the ith day.

Remark. Another solution. Let $\{1, 2, \ldots, 132\}$ be the union of 63 binary sets A_{pq} and 6 unary sets A_r, with $A_{pq} = \{42\,p + q, 42\,p + q + 21\}$, $p = 0, 1, 2, q = 1, \ldots, 21$; $A_r = \{r\}$, $r = 127, 128, \ldots, 132$. Then, put 77 numbers $a_1, a_2, \ldots, a_{77}$ into 69 *drawers*. By the drawer principle, at least two numbers lie in the same drawer A_{pq}, as we have shown.

Example 8. Suppose that the sum of 100 non-negative real numbers is 1. Prove that we can arrange them in a circle, such that the sum of 100 products of two adjacent numbers is not more than 0.01.

Solution. There are 99! circular permutations of 100 numbers, in which the number of permutations that two numbers are adjacent is $2 \times 98!$. So, the average sum of the products of two adjacent numbers over 99! permutations is

$$\frac{2 \cdot 98!}{99!} \sum_{1 \le i < j \le 100} x_i x_j = \frac{1}{99} \left(\left(\sum_{i=1}^{100} x_i \right)^2 - \sum_{i=1}^{100} x_i^2 \right)$$

$$\le \frac{1}{99} \left(\left(\sum_{i=1}^{100} x_i \right)^2 - \frac{1}{100} \left(\sum_{i=1}^{100} x_i \right)^2 \right)$$

$$= \frac{1}{100}.$$

By the average principle, there is at least one permutation such that the sum of the products of two adjacent numbers is not more than $1/100$.

Remark. The average principle can be used for non-integer cases. It is an idea from the integrated point of view.

Example 9. For a given real sequence

$$a_1, \ldots, a_{n^2+1} \tag{1}$$

with $n^2 + 1$ terms, prove that there is a monotone subsequence of (1) with $n + 1$ terms.

Solution. For each i, select the longest increasing subsequence starting from a_i, and let its length be l_i. Then we obtain a sequence of $n^2 + 1$ integers

$$l_1, \ldots, l_{n^2+1}. \tag{2}$$

If there is an $l_k \ge n + 1$, then the problem is proved. Otherwise, $0 < l_i \le n$, for $i = 1, \ldots, n^2 + 1$. By the drawer principle, in (2), there are at least $\lceil \frac{n^2+1}{n} \rceil = n + 1$ terms all equal to some positive integer l. Suppose that $l_{k_1} = l_{k_2} = \cdots = l_{k_{n+1}} = l$, $k_1 < k_2 < \cdots < k_{n+1}$. We show that the subsequence $a_{k_1}, \ldots, a_{k_{n+1}}$ is decreasing, which completes the proof. Since, otherwise, there are $k_i < k_j$, such that $a_{k_i} < a_{k_j}$. Then there is an increasing subsequence of $l + 1$ terms by adding a_{k_i} before a_{k_j}.

Remark. Example 9 is a theorem found and proved by mathematicians P. Erdös and G. Szekeres. Its simple extension is as follows: given a real sequence of $mn + 1$ terms, if every increasing subsequence has at most m terms, then there is a decreasing subsequence of $n + 1$ terms.

Example 10. In the first 10^9 terms of Fibonacci sequence $\{F_n\}$: $1, 1, 2, 3, 5, 8, \ldots, F_1 = 1$, $F_2 = 1$, $F_{n+2} = F_{n+1} + F_n$ for $n = 1, 2, \ldots$, is there a term divisible by 10,000?

Solution. The answer is positive. Denote $F_0 = 0$ for convenience. Let f_n be the remainder of F_n divided by 10,000. Then $0 \le f_n \le 9999$. By the drawer principle, in $10,000^2 + 1 = 10^8 + 1$ pairs of integers (f_n, f_{n+1}), $(0 \le n \le 10^8)$, there are identical pairs, say $(f_p, f_{p+1}) = (f_q, f_{q+1})$, $0 \le p < q \le 10^8$. Hence,

$$F_p \equiv F_q(\text{mod } 10,000), \quad F_{p+1} \equiv F_{q+1}(\text{mod } 10,000).$$

If $p = 0$, then $F_q \equiv 0(\text{mod } 10,000)$.

If $p \ge 1$, then $F_{q-1} \equiv F_{q+1} - F_q \equiv F_{p+1} - F_p \equiv F_{p-1}(\text{mod } 10,000)$, and $F_{q-2} \equiv F_{p-2}(\text{mod } 10,000), \ldots, F_{q-p+1} \equiv F_1(\text{mod } 10,000)$, $F_{q-p} \equiv F_0 \equiv 0(\text{mod } 10,000)$. So, F_{q-p} is a multiple of 10,000.

Remark. If fact, by using a computer we can obtain that $F_{7498} \equiv 4999(\text{mod } 10,000)$, $F_{7499} \equiv 5001(\text{mod } 10,000)$, so $F_{7500} \equiv 0(\text{mod } 10,000)$; $F_{14998} \equiv 9999(\text{mod } 10,000)$, $F_{14999} \equiv 1(\text{mod } 10,000)$, so $F_{15000} \equiv 0(\text{mod } 10,000)$. Thus, for any positive integer k, $F_{7500k} \equiv 0(\text{mod } 10,000)$.

Example 11. A test contains five multiple choice questions which have four options in each. Suppose each examinee choses one option for each question. There exists a number n such that for any n sheets among 2000 sheets of answer papers, there are four sheets of answer papers such that any two of them have at most three questions with the same answers. Find the minimal possible value of n (Problem from the 2000 China Mathematical Olympiad).

Solution. Denote the four choices by 1, 2, 3, and 4. And denote the answers of five problems by an array (a, b, c, d, e), where $a, b, c, d, e \in \{1, 2, 3, 4\}$.

Let the first four answers (a, b, c, d) be an array. Then there are totally $4^4 = 256$ arrays. By the drawer principle, for 2000 answer sheets, there is an array with at least $\lceil \frac{2000}{256} \rceil = 8$ sheets: take out these 8 sheets and denote them as class A. In the same way, for the remaining 1992 sheets, we can take out $\lceil \frac{1992}{256} \rceil = 8$ sheets as class B, and for the rest of the 1984 sheets, we can take out $\lceil \frac{1984}{256} \rceil = 8$ sheets as class C. Again, by the drawer principle, for any four sheets of classes A, B, and C, there are two sheets in a class. Namely, they at least have four same answers. Therefore, $n \ge 25$.

On the other hand, if $n = 25$, we construct 2000 answer sheets as follows.

Take 250 different arrays (a, b, c, d), and for each array, let $a+b+c+d+e$ be a multiple of 4. And let eight sheets have the answer (a, b, c, d, e), so, there are 2000 answer sheets in total. Take any 25 answer sheets, by the drawer principle, there are four sheets, any two of which have at most three same answers.

Exercises

1. Take any $n + 1$ numbers from $2n$ positive integers $1, 2, \ldots, 2n$. Prove that there are two coprime numbers.
2. Take any 11 numbers from 100 positive integers $1, 2, \ldots, 100$. Prove that there are two numbers whose ratio is not greater than $3/2$.
3. Given any seven numbers, prove that there are two of them, say x and y, such that $0 \le \frac{x-y}{1+xy} < \frac{1}{\sqrt{3}}$ (Problem from the 1984 Canadian Mathematical Olympiad).
4. Put six points anywhere in a rectangle of 3×4. Prove that there are two points whose distance is not greater than $\sqrt{5}$ (Problem from the 1981 All Soviet Union Mathematical Olympiad).
5. Suppose that A and B are two subsets of the set $\{1, 2, \ldots, 100\}$, satisfying $A \cap B = \emptyset$, A has the same number of elements as B, and if $n \in A$, then $2n + 2 \in B$. Find the maximum number of elements of $A \cup B$ (Problem from the 2007 China High School Mathematical League).
6. There are n students in a class taking two exams, Chinese language and maths. Each grade is an integer between 0 and 100. No two students have the same scores. Find the least integer n, such that there always exist three students, say A, B and C, such that A has higher scores than B, and B has higher scores than C (Problem of Shangdong Province preliminary from the 2011 China High School Mathematical League).
7. Fill each square with a positive integer in a grid of 50×50 squares, such that each number of $1, 2, \ldots, 50$ appears 50 times in the grid. Prove that there is a row or a column that contains at least eight different numbers.
8. Let $\{a_n\}$ be a sequence of positive integers, such that

$$a_{10k-9} + a_{10k-8} + \cdots + a_{10k} \le 19 \quad \text{for all } k \in \mathbb{N}^+.$$

Denote the sum of several consecutive terms of $\{a_n\}$ by $S(i, j) = \sum_{p=i+1}^{j} a_p$, where $j > i$, $i, j \in \mathbb{N}^+$. Prove that $\bigcup_{i<j\in\mathbb{N}^+} S(i, j) = \mathbb{N}^+$ (Problem from the 2009 Shanghai High School Mathematical Contest).

9. Suppose that the sequence $\{a_n\}$ satisfies $a_{n+3} = a_{n+2} + 2a_{n+1} + a_n$, $n \in \mathbb{N}^+$, with $a_1 = 1$, $a_2 = 1$, $a_3 = 3$. Prove that for any positive integer m, there is a positive integer n, such that $m|a_n$.

10. Suppose that each point on a circle is in one of N colours. Prove that

    (1) If $N = 2$, there is an inscribed isosceles triangle whose three vertices are in the same colour.

    (2) If $N \geq 2$, there is an inscribed trapezoid whose four vertices are in the same colour (Problem of 2007 Belarusian Mathematical Olympiad).

Chapter 5

Inclusion–Exclusion Principle

The addition principle is an important counting principle. While applying the addition principle to count the number of elements of a set, we need to divide the set into mutually disjoint subsets. However, sometimes it is difficult to find suitable subsets for counting. So, addition principle is generalized to the inclusion–exclusion principle.

We count the number of elements of every subset, with some elements to be counted more than once; then exclude the over-counted elements, and then add back to over-over-counted elements, and so on, until the exact result is obtained. That is the following basic and important

Inclusion–exclusion principle I. For n finite sets $A_1, A_2, \ldots, A_n$, we have

$$\left| \bigcup_{i=1}^{n} A_i \right| = \sum_{i=1}^{n} |A_i| - \sum_{1 \leq i < j \leq n} |A_i \cap A_j| + \sum_{1 \leq i < j < k \leq n} |A_i \cap A_j \cap A_k|$$

$$- \cdots + (-1)^{n-1} \left| \bigcap_{i=1}^{n} A_i \right|.$$

In this chapter, $|X|$ means the number of elements of set X, and $\bar{X}$ means the complement of X to some universal set I.

Note that we have generalized De Morgan's law for union and intersection: $\bigcap_{i=1}^{n} \bar{A}_i = \overline{\bigcup_{i=1}^{n} A_i}$. When the universal set I is finite, then $|\bigcap_{i=1}^{n} \bar{A}_i| = |I| - |\bigcup_{i=1}^{n} A_i|$.

Inclusion–exclusion principle II. For n finite subsets $A_1, A_2, \ldots, A_n$ of finite universal set I, we have

$$\left| \bigcap_{i=1}^{n} \bar{A}_i \right| = |I| - \sum_{i=1}^{n} |A_i| + \sum_{1 \le i < j \le n} |A_i \cap A_j| - \sum_{1 \le i < j < k \le n} |A_i \cap A_j \cap A_k|$$
$$+ \cdots + (-1)^n \left| \bigcap_{i=1}^{n} A_i \right|.$$

Example 1. How many positive integers not greater than 1000 are neither divisible by 5 nor divisible by 7? What is the sum of these numbers?

Solution. Let $I = \{1, 2, \ldots, 1000\}$, A_k be the set of multiples of k in I, then by the inclusion–exclusion principle,

$$|\bar{A}_5 \cap \bar{A}_7| = |I| - |A_5| - |A_7| + |A_{35}|$$
$$= 1000 - \left\lfloor \frac{1000}{5} \right\rfloor - \left\lfloor \frac{1000}{7} \right\rfloor + \left\lfloor \frac{1000}{35} \right\rfloor$$
$$= 1000 - 200 - 142 + 28 = 686.$$

The sum of these numbers is

$$\sum_{i=1}^{1000} i - \sum_{i=1}^{200} 5i - \sum_{i=1}^{142} 7i + \sum_{i=1}^{28} 35i$$
$$= \frac{1000 \times 1001}{2} - \frac{5 \times 200 \times 201}{2} - \frac{7 \times 142 \times 143}{2} + \frac{35 \times 28 \times 29}{2}$$
$$= 500,500 - 100,500 - 71,071 + 14,210 = 343,139.$$

Remark. To find the sum of these elements, the same idea is used.

Example 2. Suppose that $0 < a_1 < a_2 < \cdots < a_{3n-2}$, $n \in \mathbb{N}^+$. Prove that, for any two permutations $(b_1, b_2, \ldots, b_{3n-2})$ and $(c_1, c_2, \ldots, c_{3n-2})$ of $(a_1, a_2, \ldots, a_{3n-2})$, there is an $i \in \{1, 2, \ldots, 3n-2\}$, such that $a_i b_i c_i \ge a_n^3$.

Solution. Denote subsets $A = \{i | a_i \ge a_n\}$, $B = \{i | b_i \ge a_n\}$, and $C = \{i | c_i \ge a_n\}$. Then A, B, and C all have $2n - 1$ members. By the inclusion–exclusion principle, we have

$$|A \cap B| = |A| + |B| - |A \cup B| \ge 2(2n - 1) - (3n - 2) = n,$$

$$|A \cap B \cap C| = |A \cap B| + |C| - |(A \cap B) \cup C| \ge n + (2n - 1) - (3n - 2) = 1.$$

Therefore, there is an i, $i \in A \cap B \cap C$, and $a_i b_i c_i \ge a_n^3$.

Example 3. Let $(a_1, a_2, \ldots, a_n)$ be a permutation of $(1, 2, \ldots, n)$. If $a_i \neq i$ for all $i = 1, 2, \ldots, n$, we say such a permutation is a *derangement* of $(1, 2, \ldots, n)$. Find the number D_n of all *derangements* of $(1, 2, \ldots, n)$ (see also Example 4 of Chapter 13).

Solution. Let I be the set of all permutations of $(1, 2, \ldots, n)$, then $|I| = n!$. Denote $A_i = \{(a_1, a_2, \ldots, a_n) \in I | a_i = i\}$, $i = 1, 2, \ldots, n$, then $|A_i| = (n-1)!$.

And similarly, $|\bigcap_{1 \leq i_1 < i_2 < \cdots < i_s \leq n} A_i| = (n-s)!$ for $s = 2, 3, \ldots, n$.

Obviously, $D_n = |\bigcap_{i=1}^{n} \bar{A}_i|$. By the inclusion–exclusion principle, we have

$$D_n = |I| - \sum_{i=1}^{n} |A_i| + \sum_{1 \leq i < j \leq n} |A_i \cap A_j| - \cdots + (-1)^n \left| \bigcap_{i=1}^{n} A_i \right|$$

$$= \sum_{k=0}^{n} (-1)^k \binom{n}{k} (n-k)! = n! \sum_{k=0}^{n} \frac{(-1)^k}{k!}.$$

Example 4. Find the number of permutations $(a_1, a_2, a_3, a_4, a_5)$ of $(1, 2, 3, 4, 5)$, such that $(a_1, a_2, \ldots, a_i)$ is not a permutation of $(1, 2, \ldots, i)$, for $i = 1, 2, 3, 4$.

Solution. Let A_i be the set of permutations of $(1, 2, 3, 4, 5)$, for which the first i number(s) is a permutation of $(1, 2, \ldots, i)$, $i = 1, 2, 3, 4$. Then,

$$|A_1| = 4! = 24, \quad |A_2| = 2! \times 3! = 12, \quad |A_3| = 3! \times 2! = 12,$$

$$|A_4| = 4! = 24, \quad |A_1 \cap A_2| = 3! = 6, \quad |A_1 \cap A_3| = 2! \times 2! = 4,$$

$$|A_1 \cap A_4| = 3! = 6, \quad |A_2 \cap A_3| = 2! \times 2! = 4, \quad |A_2 \cap A_4| = 2! \times 2! = 4,$$

$$|A_3 \cap A_4| = 3! = 6, \quad |A_1 \cap A_2 \cap A_3| = 2! = 2, \quad |A_1 \cap A_2 \cap A_4| = 2! = 2,$$

$$|A_1 \cap A_3 \cap A_4| = 2! = 2, \quad |A_2 \cap A_3 \cap A_4| = 2! = 2, \quad |A_1 \cap A_2 \cap A_3 \cap A_4| = 1.$$

Thus, by the inclusion–exclusion principle,

$$|A_1 \cup A_2 \cup A_3 \cup A_4|$$

$$= |A_1| + |A_2| + |A_3| + |A_4| - |A_1 \cap A_2| - |A_1 \cap A_3| - |A_1 \cap A_4|$$

$$- |A_2 \cap A_3| - |A_2 \cap A_4| - |A_3 \cap A_4| + |A_1 \cap A_2 \cap A_3| + |A_1 \cap A_2 \cap A_4|$$

$$+ |A_1 \cap A_3 \cap A_4| + |A_2 \cap A_3 \cap A_4| - |A_1 \cap A_2 \cap A_3 \cap A_4|$$

$$= 24 + 12 + 12 + 24 - 6 - 4 - 6 - 4 - 4 - 6 + 2 + 2 + 2 + 2 - 1 = 49.$$

Therefore, the number of desired permutations is $5! - 49 = 71$.

Example 5. An association with n members in total consists of six committees. Each committee has at least $n/4$ members. Prove that there are two committees with $n/30$ or more members in common.

Solution. Let A_i, $i = 1, 2, 3, 4, 5, 6$ be the membership of the committees, respectively. Then $n = |\bigcup_{i=1}^{6} A_i| \geq \sum_{i=1}^{6} |A_i| - \sum_{1 \leq i < j \leq 6} |A_i \cap A_j|$, so, $\sum_{1 \leq i < j \leq 6} |A_i \cap A_j| \geq \sum_{i=1}^{6} |A_i| - n \geq 6 \times \frac{n}{4} - n = \frac{n}{2}$.

Hence, there are $1 \leq i < j \leq 6$, such that

$$|A_i \cap A_j| \geq \frac{1}{\binom{6}{2}} \times \frac{n}{2} = \frac{n}{30}.$$

Example 6. A permutation $(x_1, x_2, \ldots, x_{2n})$ of the set $\{1, 2, \ldots, 2n\}$, where n is a positive integer, is said to have the property P if $|x_i - x_{i+1}| = n$ for at least one i in $\{1, 2, \ldots, 2n-1\}$. Show that, for each n, there are more permutations with property P than without (Problem from the 1989 International Mathematical Olympiad).

Solution. Let A_k be all the permutations of $\{1, 2, \ldots, 2n\}$ with k and $k + n$ being adjacent, $k = 1, 2, \ldots, n$. Then, $A = \bigcup_{k=1}^{n} A_k$ is the set of permutations with the property P. By the inclusion–exclusion principle,

$$|A| = \left| \bigcup_{i=1}^{n} A_i \right| \geq \sum_{i=1}^{n} |A_i| - \sum_{1 \leq i < j \leq n} |A_i \cap A_j|. \tag{1}$$

Note that $|A_i| = 2 \times (2n-1)!$; $|A_i \cap A_j| = 2^2 \times (2n-2)!$, since there are two permutations for each of $(i, i+n)$ and $(j, j+n)$; then we consider each pair as one 'number', the two 'numbers' together with the rest $2n - 4$ numbers have $(2n - 2)!$ permutations. Thus,

$$|A| \geq 2n(2n-1)! - 4 \binom{n}{2}(2n-2)! = \frac{n(2n)!}{2n-1} > \frac{(2n)!}{2},$$

which completes the proof.

Remark. We can use (1) to give a lower bound for $|A|$, and further by the inclusion–exclusion principle we can obtain an exact summation for $|A|$:

$$|A| = \sum_{k=1}^{n} (-1)^{k-1} 2^k (2n - k)! \binom{n}{k}.$$

Example 7. Prove that, for n finite sets $A_1, A_2, \ldots, A_n$, we have

$$|A_1 \cap A_2 \cap \cdots \cap A_n|$$

$$= \sum_{i=1}^{n} |A_i| - \sum_{1 \le i < j \le n} |A_i \cup A_j| + \sum_{1 \le i < j < k \le n} |A_i \cup A_j \cup A_k|$$

$$- \cdots + (-1)^{n-1} |A_1 \cup A_2 \cup \cdots \cup A_n|. \tag{1}$$

Solution. Take a finite set I containing $A_1 \cup A_2 \cup \cdots \cup A_n$ as the universal set.

Note that A_i and $\bar{A}_i$ are complementary, by the inclusion–exclusion principle,

$$|A_1 \cap A_2 \cap \cdots \cap A_n| = |I| - \sum_{i=1}^{n} |\bar{A}_i| + \sum_{1 \le i < j \le n} |\bar{A}_i \cap \bar{A}_j|$$

$$- \sum_{1 \le i < j < k \le n} |\bar{A}_i \cap \bar{A}_j \cap \bar{A}_k|$$

$$+ \cdots + (-1)^n |\bar{A}_1 \cap \bar{A}_2 \cap \cdots \cap \bar{A}_n|. \tag{2}$$

For any $1 \le i_1 < i_2 < \cdots < i_s \le n$, we have

$$|\bar{A}_{i_1} \cap \bar{A}_{i_2} \cap \cdots \cap \bar{A}_{i_s}| = |\overline{A_{i_1} \cup A_{i_2} \cup \cdots \cup A_{i_s}}| = |I| - |A_{i_1} \cup A_{i_2} \cup \cdots \cup A_{i_s}|,$$

summing up for $i_1, i_2, \ldots, i_s$, to yield

$$\sum_{i_1, i_2, \ldots, i_s} |\bar{A}_{i_1} \cap \bar{A}_{i_2} \cap \cdots \cap \bar{A}_{i_s}| = \binom{n}{s} \cdot |I| - \sum_{i_1, i_2, \ldots, i_s} |A_{i_1} \cup A_{i_2} \cup \cdots \cup A_{i_s}|. \tag{3}$$

Denote the right-hand side of (1) by S. By (1) and (2), we have

$$|A_1 \cap A_2 \cap \cdots \cap A_n|$$

$$= |I| - \left(n \cdot |I| - \sum_{i=1}^{n} |A_i| \right) + \left(\binom{n}{2} \cdot |I| - \sum_{1 \le i < j \le n} |A_i \cap A_j| \right)$$

$$- \left(\binom{n}{3} \cdot |I| - \sum_{1 \le i < j < k \le n} |A_i \cap A_j \cap A_k| \right)$$

$$+ \cdots + (-1)^n \left(\binom{n}{n} \cdot |I| - |A_1 \cap A_2 \cap \cdots \cap A_n| \right)$$

$$= |I| \left(\binom{n}{0} - \binom{n}{1} + \binom{n}{2} - \cdots + (-1)^n \binom{n}{n} \right) + S = S.$$

This completes the proof.

At the end of this chapter, we present an example that requires high skills.

Example 8. n finite sets $A_1, A_2, \ldots, A_n$, satisfy

$$|A_k \cap A_{k+1}| > \frac{n-2}{n-1}|A_{k+1}|, \quad k = 1, 2, \ldots, n(A_{n-1} = A_1). \tag{1}$$

Prove that $\bigcap_{i=1}^n A_i \neq \varnothing$ (A training problem from the 2007 China Team).

Solution. For $k = 2, 3, \ldots, n$, let $B_k = \bigcap_{i=1}^k A_i$. We show that, for $k = 2, 3, \ldots, n$,

$$|B_k| > \frac{n-2}{n-1}|A_k| - \frac{1}{n-1}\sum_{i=2}^{k-1}|A_i|. \tag{2}$$

We prove (2) by induction on k. If $k = 2$, $|B_2| = |A_1 \cap A_2| > \frac{n-2}{n-1}|A_2|$, (2) is true.

Suppose that (2) is true for $n > k \geq 2$. For $k + 1$, by the inclusion–exclusion principle,

$$|B_k \cup (A_k \cap A_{k+1})| = |B_k| + |A_k \cap A_{k+1}| - |B_k \cap (A_k \cap A_{k+1})|.$$

So, we obtain

$$|B_{k+1}| = |B_k \cap (A_k \cap A_{k+1})| = |B_k| + |A_k \cap A_{k+1}| - |B_k \cup (A_k \cap A_{k+1})|.$$

Obviously, $B_k, A_k \cap A_{k+1} \subseteq A_k$, so $B_k \cup (A_k \cap A_{k+1}) \subseteq A_k$, and

$$|B_k \cup (A_k \cap A_{k+1})| \leq |A_k|.$$

Therefore, by inductive assumption and (2), we have

$$|B_{k+1}| \geq |B_k| + |A_k \cap A_{k+1}| - |A_k|$$

$$> \frac{n-2}{n-1}|A_k| - \frac{1}{n-1}\sum_{i=2}^{k-1}|A_i| + \frac{n-2}{n-1}|A_{k+1}| - |A_k|$$

$$= \frac{n-2}{n-1}|A_{k+1}| - \frac{1}{n-1}\sum_{i=2}^{k}|A_i|.$$

That is, (2) is true for $k + 1$, which completes the induction.

Especially, (2) is true for $k = n$:

$$|B_n| > \frac{n-2}{n-1}|A_n| - \frac{1}{n-1}\sum_{i=2}^{n-1}|A_i|. \tag{3}$$

Without loss of generality, we may assume that $|A_n| = \max_{1 \le i \le n}|A_i|$, then by (3), $|B_n| > 0$, and $\bigcap_{i=1}^{n} A_i = B_n \ne \varnothing$.

Exercises

1. An $n(n \ge 3)$ digits number consists all of 1, 2, and 3 only. Find the number of such integers.

2. In the set of all positive integers $\{1, 2, 3, 4, 5, \ldots\}$, delete multiples of 3 and multiples of 4, but keep the multiples of 5 (e.g. 15, 20, ...). Denote the remaining numbers in the increasing order by a sequence $\{a_n\}$. Find a_{2005}.

3. Denote the number of all subsets of finite set S by $n(S)$. If A, B, and C are finite sets, and $n(A) + n(B) + n(C) = n(A \cup B \cup C)$, $|A| = 100 = |B|$. Find the minimum of $|A \cap B \cap C|$.

4. If a cuboid of $a \times b \times c$ consists of abc unit cubes, $(a, b, c \in \mathbb{N}^+)$, find the number of cubes that are crossed by a diagonal of the cuboid.

5. In a region with area 1001, there are 2001 subregions S_i $(1 \le i \le 2001)$ each with area 1. Prove that, there are two subregions, S_i and S_j, such that the area of their common part is not less than $1/2001$.

6. In a live-fire drill, the blue army plans to supply five different new weapons on a line of defense with 20 posts, such that there is at least one weapon for every five adjacent posts, but not for the first and the last posts, and not for two adjacent posts simultaneously. How many supply schemes are there? (Problem from the 2005 Zhejiang Province Preliminary for China High School Mathematical League).

7. Let $(A_1, \ldots, A_m)$ be a group of m nonempty sets, such that $\bigcup_{i=1}^{m} A_i = \{1, 2, \ldots, n\}$. Denote the number of all such groups by $f(m, n)$. Prove that $f(m, n) = \sum_{k=0}^{m-1}(-1)^k \binom{m}{k}(2^{m-k} - 1)^n$.

8. Let $a, b, c > 1$ satisfy $\frac{1}{a} + \frac{1}{b} + \frac{1}{c} > 1$. Define set $S(x) = \{\lfloor kx \rfloor | k \in \mathbb{N}^+\}$. Prove that there must be two of three sets $S(a)$, $S(b)$, and $S(c)$, that have infinite intersection.

Chapter 6

Extreme Principle

Extreme principle is a method of considering an object's special properties, such as the maximum, the minimum, the longest, or the shortest, to solve problems. It has wide applications in geometry, number theory, combinatorics, and graph theory. By using this simple and powerful principle, we can solve mathematical problems, especially the existence problems. The method varies when applied to different problems.

In applying the principle, the following **Facts** (1) and (2) are very important and we should pay attention to (3) as well:

(1) In a finite set of real numbers, there is a maximum and a minimum.
(2) In a set of infinitely many integers with lower bound, there is a minimum.
(3) In a set of infinitely many real numbers, there is not necessarily a maximum or a minimum.

Example 1. Suppose there are $2n + 1(n \in \mathbb{N}^+)$ planets with mutually different distances. People of each planet have observed the nearest planet. Prove that, there is only one planet that has not been observed.

Solution. There is a unique pair of planets that are the nearest, they have been observed by each other, but have not observed others. So, we can ignore these two planets and consider other $(2n-1)$ planets. We repeatedly apply this until one planet is left. That planet has not been observed.

Remark. In this problem, the extreme is the shortest distance.

In general, when carrying out induction on n, the element with certain extreme properties can be removed so as to reduce the case by $n - 1$, or smaller.

Example 2. Colour $n(\geq 5)$ points on the plane red or blue. If any three points in the same colour are not collinear, prove that there is a triangle such that:

(1) Its three vertices are in the same colour.
(2) There is a side on which there is no points in the other colour.

Solution. Since the number of points is $n \geq 5$, and there are at most two colours. We see that at least three points are in the same colour. Take any three of them to form a triangle. So (1) is proved.

Take one triangle in (1) with the smallest area. Then this triangle satisfies (2) since if each side of this triangle contains a point in the other color, then there is a smaller triangle whose vertices are of the same colour.

Remark. In this problem, the extreme is the smallest area.

Example 3. Prove that the Diophantine equation $x^3 + 2y^3 = 4z^3$ has no solution (x, y, z) of positive integers.

Solution. We prove by contradiction. Suppose, on the contrary, there is a solution (x_1, y_1, z_1) of positive integers with x_1 being the smallest, so,

$$x_1^3 + 2y_1^3 = 4z_1^3.$$

We see that x_1^3 is even, and so is x_1. Let $x_1 = 2x_2$, then

$$8x_2^3 + 2y_1^3 = 4z_1^3,$$

that is,

$$4x_2^3 + y_1^3 = 2z_1^3.$$

Thus, y_1 is even. Let $y_1 = 2y_2$, then

$$4x_2^3 + 8y_2^3 = 2z_1^3.$$

That is, $2x_2^3 + 4y_2^3 = z_1^3$.
So, z_1 is even. Let $z_1 = 2z_2$, then

$$2x_2^3 + 4y_2^3 = 8z_2^3.$$

Hence, (x_2, y_2, z_2) is also a solution of positive integers with $x_2 < x_1$. A contradiction.

Example 4. Given any $m \times n$ matrix of real numbers

$$\begin{pmatrix} a_{11} & a_{12} & \cdots & a_{1n} \\ a_{21} & a_{22} & \cdots & a_{2n} \\ \cdots & \cdots & \cdots & \cdots \\ a_{m1} & a_{m2} & \cdots & a_{mn} \end{pmatrix}.$$

An operation on the matrix is to change the signs of all numbers on one row or on one column, and leave all other numbers unchanged. Prove that we can perform finitely many operations on the matrix, such that the sum of each row and the sum of each column are non-negative.

Solution. First, we note that each element of the matrix takes at most two values. Thus, the sum S of all elements of the matrix takes at most 2^{mn} possible values. So, there is a maximum of S.

Take a matrix with the maximum S, and we show that this matrix satisfies the required conditions.

Otherwise, suppose that the sum of numbers on the ith row (column) is less than 0, Take an operation on this row (column), then the sum of all elements of the matrix will be larger than S. A contradiction.

Example 5. Prove that in every tetrahedron there is a vertex such that the three edges meeting there have lengths which are the sides of a triangle (Problem from the 1968 International Mathematical Olympiad).

Solution. Let the tetrahedron be $ABCD$ with the longest edge AB. We show that three edges incident to vertex A or B can be the sides of a triangle.

In fact, for $\triangle ABC$ and $\triangle ABD$, we have

$$AC + CB > AB, \quad AD + DB > AB.$$

Summing up the inequalities, we yield

$$AC + CB + AD + DB > 2AB.$$

Rearrange the terms to obtain

$$(AC + AD) + (BC + BD) > 2AB.$$

Hence, $AC + AD > AB$ or $BC + BD > AB$. If $AC + AD > AB$, since AB is the longest, then edges AB, AC, and AD can be three sides of a triangle. If $BC + BD > AB$, edges BC, BD, and BA can be three sides of a triangle.

Remark. In the above problem, the extreme is the longest side of a triangle. So, the conditions that three edges form a triangle are reduced to one condition.

Example 6. Let sets $A_1, A_2, \ldots, A_n$ and sets $B_1, B_2, \ldots, B_n$ be two partitions of set M, such that if any two sets A_i and B_j are disjoint, then $|A_i \cup B_j| \geq n$. Prove that $|M| \geq \frac{n^2}{2}$.

Solution. Without loss of generality, suppose that A_1 has the least number of elements among $A_i, B_j (1 \leq i \leq n, \ 1 \leq j \leq n)$ and $|A_1| = p$.

If $p \geq \frac{n}{2}$, then,

$$|M| = |A_1| + |A_2| + \cdots + |A_n| \geq np \geq \frac{n^2}{2}.$$

If $p < \frac{n}{2}$, divide $B_1, B_2, \ldots, B_n$ into two classes, such that those having common members with A_1 are of class one, say $B_1, B_2, \ldots, B_q$ with $q \leq p$, and those that do not are of class two, say $B_{q+1}, B_{q+2}, \ldots, B_n$, which all have at least $n - p$ elements by $|A_1 \cup B_j| \geq n, j = q + 1, \ldots, n$.

Thus,

$$|M| = (|B_1| + \cdots + |B_q|) + (|B_{q+1}| + \cdots + |B_n|)$$
$$\geq qp + (n - q)(n - p)$$
$$= \frac{n^2}{2} + \frac{1}{2}(n - 2p)(n - 2q).$$

Since $q \leq p < n/2$, we have $(n - 2p)(n - 2q) > 0$, therefore, $|M| > \frac{n^2}{2}$.

Remark. The sixth problem of the 1971 International Mathematical Olympiad is as follows:

Let $A = (a_{ij})(i, j = 1, \ldots, n)$ be a square matrix whose elements are non-negative integers. Suppose that whenever an element $a_{ij} = 0$, the sum of the elements in the ith row and the jth column is $\geq n$. Prove that the sum of all the elements of the matrix is $\geq n^2/2$.

This problem can be solved by a similar method as in the solution of Example 6.

Example 7. There are $n(\geq 7)$ circles on a plane and any three of them have no pairwise intersections. Prove that there is a circle that intersects with most of the other five circles.

Solution. Let the centres of n circles be $O_1, O_2, \ldots,$ and O_n, respectively. Let the radius of $\odot O_1$ be the least. We show that $\odot O_1$ intersects

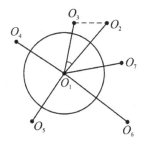

Figure 6.1

with at most 5 circles. Otherwise, if $\odot O_1$ intersects six circles, say, the centres O_2, $O_3, \ldots$, and O_7 are in the anticlockwise order. Connect $O_1O_2, O_1O_3, \ldots, O_1O_7$ as shown in Fig. 6.1. Then there is an angle in $\angle O_2O_1O_3$, $\angle O_3O_1O_4$, $\angle O_4O_1O_5$, $\angle O_5O_1O_6$, $\angle O_6O_1O_7$, $\angle O_7O_1O_2$ that is not greater than $\frac{360°}{6} = 60°$, say, $\angle O_2O_1O_3 \leq 60°$.

Connect O_2O_3, and denote the radii of $\odot O_1$, $\odot O_2$, $\odot O_3$ by r_1, r_2, r_3, respectively. Then $r_1 \leq r_2$, $r_1 \leq r_3$.

Since $\odot O_1$ intersects with both $\odot O_2$ and $\odot O_3$, we have $r_1 + r_2 \geq O_1O_2$, $r_1 + r_3 \geq O_1O_3$, thus,

$$O_1O_2 \leq r_1 + r_2 \leq r_2 + r_3,$$
$$O_1O_3 \leq r_1 + r_3 \leq r_2 + r_3.$$

But $\angle O_2O_1O_3 \leq 60°$, then there must be an angle $\geq 60°$ in $\triangle O_1O_2O_3$, say, $\angle O_3 \geq 60°$. Then $O_1O_2 \geq O_2O_3$, so,

$$O_2O_3 \leq r_1 + r_2 \leq r_2 + r_3.$$

Hence, $\odot O_2$ and $\odot O_3$ intersect. And $\odot O_1$, $\odot O_2$, $\odot O_3$ have pairwise intersections. A contradiction.

Example 8. Determine all possible values of integer k, for which there exist positive integers a and b such that $\frac{b+1}{a} + \frac{a+1}{b} = k$ (Problem from the 2010 Western China Mathematical Olympiad).

Solution. For fixed positive integer $k > 2$, let $b \geq 1$ be the smallest positive integer such that $\frac{b+1}{a} + \frac{a+1}{b} = k$. Then by the symmetry of a and b, we have $a \geq b \geq 1$. We see that $x = a$ is a root of the quadratic equation $\frac{b+1}{x} + \frac{x+1}{b} = k$, let the other root be a'. By Viète's theorem, we have,

$a + a' = kb - 1 > 1$ and $a \cdot a' = b(b+1)$. So, $(a, b) = (b, b)$ or $(b + 1, b)$. Thus, $k = 2 + \frac{2}{b}$. Taking $b = 1, 2$, we get the answer: $k = 3, 4$.

Example 9. Given $n(n \geq 3)$ points on the plane that not all on a line. Prove that there is a line passing through only two points of these n points.

Solution. For each line passing through at least two points of the given n points, there is a point not on the line that has the shortest distance to the line. In these finite distances, there is a shortest one, say, the distance d_0 from point P to the line AB. We shall prove that the line AB passes through exactly two points.

We prove by contradiction. Let $PH = d_0$, H be the perpendicular foot point on the line AB. If there is another point C on AB, then of A, B, C, there are at least two points on the same side of H, say, they are B and C and C is nearer to H (see Fig. 6.2). Then the distance from C to line PB, is shorter than $PH = d_0$, a contradiction.

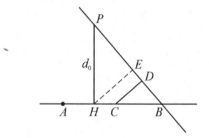

Figure 6.2

Remark. This problem was posted and named by English mathematician J. J. Sylvester (1814–1897). It seems simple but was not solved in his lifetime. Afterword, many mathematicians attempted but failed. Such situation lasted for 50 years.

A problem dual to Sylvester's problem is

> On a plane, given n lines not pairwisely parallel. If at intersection point of any two lines there is another line passing through it. Prove that these n lines intersect at a point. The proof is similar and we leave it to the reader.

Example 10. Let f be a function defined on non-negative real numbers with $f(0) = 0$, satisfying

$$|f(x) - f(y)| \leq (x - y)f(x), \quad \forall x \geq y \geq 0. \tag{1}$$

Find f.

Solution. (see Example 9 of Chapter 3) Here, we give a solution for you to judge.

We show that $f(x) \equiv 0$.

Suppose on the contrary, there is a least $x_0 > 0$ such that $f(x_0) \neq 0$.

Let $x = x_0, y = \max\left\{\frac{x_0}{2}, x_0 - \frac{1}{2}\right\}$, then $x > y \geq 0$ and $x - y \leq \frac{1}{2}$. Since x_0 is the least, $f(y) = 0$. Substitute $x = x_0$ to (1), we have,

$$|f(x_0)| \leq (x_0 - y)f(x_0) \leq \frac{1}{2}|f(x_0)| < |f(x_0)|.$$

A contradiction!

Anything wrong? If we change the initial condition $f(0) = 0$ to $f(0) = 1$, and follow the above way, we shall obtain that $f(x) \equiv 1$, by taking $x = x_0, y = \max\left\{\frac{x_0}{2}, x_0 - \frac{|f(x_0)-1|}{|f(x_0)|+1}\right\}$. However, with a little knowledge of calculus, we know that $f(x) = e^x$ is the right answer.

What is wrong with it?

In fact, for infinitely many real numbers with lower bounds, the least number might not exist.

Of course, for the original problem, we can use extreme principle on intervals $\left(\frac{n-1}{2}, \frac{n}{2}\right]$ ($n \in \mathbb{N}^+$), and solve the problem correctly.

Exercises

1. In a round robin tournament of table tennis of n (≥ 3) players, no player wins all the games. Prove that there are three players A, B, and C, such that A beats B, B beats C, and C beats A.

2. Given 100 points on a plane, where the distance of any two points is not greater than 1, and the triangle formed by any three points is obtuse, prove that these 100 points can be covered by a disc with radius $1/2$.

3. There are eight points in space, where any four of them are not on a plane. Connect 17 line segments between these points. Prove that they form as least one triangle.

4. Several people gather at a party, and some are familiar with each other. If two people have the same number of familiars at the party, then they have no familiars in common. Prove that if someone has at least 2012 familiars at the party, then there must be someone who has exactly 2012 familiars at the party.

5. Of all perpendicular lines drawn from a point in a convex polygon to its sides, prove that there is at least one whose foot is on the side (but not at the vertex).

6. In a country, there are some cities linked together by roads. The roads just meet each other inside the cities. In each city, there is a board which shows the shortest length of the road originating in that city and going through all other cities (the way can go through some cities more than one time and it is not necessary to turn back to the originated city). Prove that the ratio of two numbers on the boards of any two cities cannot be greater than 3/2 or less than 2/3 (Problem from the 2009 All-Russian Mathematical Olympiad).

7. Prove that there is no regular n-gon ($n \geq 7$) with lattice vertices in the rectangular coordinate plane.

8. Put 2007 integers on a circle, such that in any adjacent five numbers there are three numbers whose sum is twice the sum of the other two numbers. Prove that all numbers are 0 (Problem from the 2007 Belarusian Mathematical Olympiad).

9. Let n be an integer greater than or equal to 2. Prove that if $k^2 + k + n$ is prime for all integers k such that $0 \leq k \leq \sqrt{n/3}$, then $k^2 + k + n$ is prime for all integers k such that $0 \leq k \leq n - 2$ (Problem from the 1987 International Mathematical Olympiad).

Chapter 7

Parity

The set of integers can be divided into sets of odd numbers and even numbers. The basic parity properties are as follows:

Property 1. An odd number is not equal to an even number.

Property 2. The sum and the difference of two integers have the same parity.

Property 3. $m \pm n$ is even if and only if m and n have the same parity; $m \pm n$ is odd if and only if m and n have different parities.

Property 4. The sum of odd number of odd numbers is odd. The sum of even number of odd numbers is even.

Property 5. An integer is odd if and only if its factors are all odd.

Property 6. Any positive integer n can be expressed in the form of $n = 2^p \cdot q$, where $p \in \mathbb{N}$ and q is odd.

Parity is the most fundamental property of integers. So, parity can be used widely and involves many important ideas and techniques.

Example 1. Let $a_1, a_2, \ldots, a_n \in \mathbb{Z}$, where n is a positive odd number, $(b_1, b_2, \ldots, b_n)$ is a permutation of $(a_1, a_2, \ldots, a_n)$. Prove that $\prod_{i=1}^{n} (a_i + |b_i|)$ and $\prod_{i=1}^{n} (a_i - |b_i|)$ are all even.

Solution. For integer x, we have $|x| \equiv x \equiv -x \,(\mathrm{mod}\, 2)$, thus

$$\sum_{i=1}^{n} (a_i + |b_i|) \equiv \sum_{i=1}^{n} (a_i - b_i) \equiv \sum_{i=1}^{n} a_i - \sum_{i=1}^{n} b_i \equiv 0 \,(\mathrm{mod}\, 2).$$

Since the left-hand side of above is the sum of an odd number of terms, there is an even term $a_i + |b_i|$ $(1 \leq i \leq n)$, hence $\prod_{i=1}^{n}(a_i + |b_i|)$ is even. Similarly, $\prod_{i=1}^{n}(a_i - |b_i|)$ is even.

Remark. Take parity as a classification by modulo 2, then, changing the operation of taking absolute value into identity operation, the operation of plus and minus are the same in modulo 2, and turn multiplication to logical operation in some sense. These facts provide great convenience to parity analysis.

Example 2. Each of the numbers $x_1, x_2, \ldots, x_n (n \geq 4)$ equals 1 or -1 and

$$\sum_{k=1}^{n} x_k x_{k+1} x_{k+2} x_{k+3} = 0, \tag{1}$$

where $x_{n+i} = x_i$ for all i. Prove that $4 \mid n$ (Problem of 1985 International Mathematical Olympiad Longlist).

Solution. Since $x_k x_{k+1} x_{k+2} x_{k+3} \in \{1, -1\}$, for $k = 1, 2, \ldots, n$, we see that the number of terms with value 1 equals the number of terms with value -1, so $n = 2m$ is even.

Since $\prod_{k=1}^{n} x_k x_{k+1} x_{k+2} x_{k+3} = (x_1 x_2 \ldots x_n)^4 = 1$, and the value on the left-hand side of the expression is $1^m \cdot (-1)^m = (-1)^m$, we see that m is even. Therefore, $n = 2m$ is a multiple of 4.

Example 3. Let a and b be two positive integers satisfying

$$(1111 + 1a)(1111 - 1b) = 123456789. \tag{1}$$

Prove that $a - b$ is a multiple of 4.

Solution. By (1), $11111 + a$ and $11111 - b$ are both odd. So, a and b are all even. Expanding (1) and simplifying, we have

$$11111(a - b) = ab + 2468.$$

Since ab is a multiple of 4, and so is $2468 = 4 \times 617$, then $11111 \times (a - b)$ is a multiple of 4, so is $a - b$.

Example 4. Select 10 elements from the set $\{0, 1, 2, \ldots, 13, 14\}$ and fill in the circles of Fig. 7.1, one number in a circle. Such that the absolute value of the difference between every pair of numbers in adjacent circle is different from each other. Can it be done? Prove your conclusion.

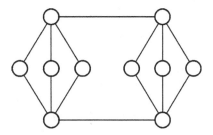

Figure 7.1

Solution. The answer is negative.

Otherwise, there are 14 distinct absolute values of differences that are not greater than 14 and not less than 1. Then they must be $1, 2, 3, \ldots, 14$, and their sum is

$$S = 1 + 2 + \cdots + 14 = 7 \times 15 = 105,$$

which is odd.

On the other hand, each circle is connected to an even number of circles. If the number filled in a circle is a, then a appears in S for even times (2 or 4). So, the result of plusor minus of even number of a is even. Thus, S is even. That is, S is odd and even as well, a contradiction.

Example 5. In a plane rectangular coordinate system, there are n lattice points, any three of which are not collinear. If the area of the triangle formed by any three points is not an integer, find the maximal possible value of n.

Solution. Divide lattice points into four sets:

$$M_1 = \{(x, y)|x \equiv y \equiv 0 \,(\mathrm{mod}\,2)\},$$

$$M_2 = \{(x, y)|x \equiv 1 \,(\mathrm{mod}\,2), y \equiv 0 \,(\mathrm{mod}\,2)\},$$

$$M_3 = \{(x, y)|x \equiv 0 \,(\mathrm{mod}\,2), y \equiv 1 \,(\mathrm{mod}\,2)\},$$

$$M_4 = \{(x, y)|x \equiv y \equiv 1 \,(\mathrm{mod}\,2)\}.$$

If $n \geq 5$, by the drawer principle, there are two points $A(x_1, y_1), B(x_2, y_2)$ in a set, $x_1 \equiv x_2 \,(\mathrm{mod}\,2), y_1 \equiv y_2 \,(\mathrm{mod}\,2)$. Then, the midpoint M of AB is a lattice point.

Take any other point $C(x_3,\ y_3)$, then

$$x_1(y_2 - y_3) + x_2(y_3 - y_1) + x_3(y_1 - y_2)$$
$$\equiv x_1(y_1 - y_3) + x_1(y_3 - y_1) + 0 \equiv 0 \,(\mathrm{mod}\,2).$$

Thus, the area of $\triangle ABC$,

$$S_{\triangle ABC} = \frac{1}{2}\,|x_1(y_2 - y_3) + x_2(y_3 - y_1) + x_3(y_1 - y_2)|$$

is an integer, a contradiction. So, $n \le 4$.

On the other hand, take four points: $(0,0), (1,0), (0,1), (1,1)$, it is easy to check that the area of any triangle is not an integer.

So, the maximal possible value of n is 4.

Remark. The following are two harder problems for the reader:

In a plane rectangular coordinate system, there are n lattice points, any three of which are not collinear.

(1) If the area of the triangle formed by any three points is odd, find the maximal possible value of n.
(2) If the area of the triangle formed by any three points is not even, find the maximal possible value of n.

Example 6. Let $n \in \mathbb{N}^+$ such that $37.5^n + 26.5^n$ is a positive integer. Find the value(s) of n (Problem of Shanghai High School Mathematical Contest).

Solution. It is easy to see that $37.5^n + 26.5^n = \frac{1}{2^n}(75^n + 53^n)$.

If n is even, then

$$75^n + 53^n \equiv (-1)^n + 1^n \equiv 2 \,(\mathrm{mod}\,4).$$

That is, $75^n + 53^n = 4m + 2$, where $m \in \mathbb{N}^+$.

Thus, $37.5^n + 26.5^n = \frac{1}{2^{n-1}}(2m + 1)$ is not an integer.

If n is odd, then

$$75^n + 53^n = (75 + 53)(75^{n-1} - 75^{n-2} \cdot 53 + \cdots + 53^{n-1})$$
$$= 2^7 \cdot (75^{n-1} - 75^{n-2} \cdot 53 + \cdots + 53^{n-1}).$$

There are n terms of odd numbers in the parenthesis, so the sum is odd. Therefore, only when $n = 1, 3, 5$ or 7, $37.5^n + 26.5^n$ is a positive integer.

Example 7. Let a, b, c, d be odd integers such that $0 < a < b < c < d$ and $ad = bc$. Prove that if $a + d = 2^k$ and $b + c = 2^m$ for some integers k and m, then $a = 1$ (Problem of 1984 International Mathematical Olympiad).

Solution. Since $ad = bc$ and $d - a > c - b > 0$, we have

$$(a + d)^2 = (d - a)^2 + 4ad > (c - b)^2 + 4bc = (b + c)^2.$$

Hence, $k > m$. Further, since

$$a(2^k - a) = ad = bc = b(2^m - b),$$

we have $2^m b - 2^k a = b^2 - a^2$, that is,

$$2^m(b - 2^{k-m}a) = (b + a)(b - a). \tag{1}$$

Note that a and b are odd, so $b + a$ and $b - a$ are even, and the difference of them is $2a \equiv 2 \pmod 4$, thus one of them can be divided by 2 but not by 4. Note that $b - 2^{k-m}a$ is odd, therefore there are only two cases:

$$\begin{cases} b + a = 2^{m-1}u, \\ b - a = 2v \end{cases} \quad \text{or} \quad \begin{cases} b + a = 2v, \\ b - a = 2^{m-1}u, \end{cases} \quad u, v \in \mathbb{N}^+. \tag{2}$$

In the first case, $uv = 2^{-m}(b^2 - a^2) = b - 2^{k-m}a < b - a = 2v$, thus $u = 1$.

Since

$$b - a < b < \frac{b + c}{2} = 2^{m-1},$$

we see that the second case cannot happen. Hence, by (1), we have

$$\begin{cases} b + a = 2^{m-1}, \\ b - a = 2(b - 2^{k-m}a). \end{cases}$$

Eliminate b to yield $a = 2^{2m-2-k}$. Since a is odd, we see that $a = 1$.

Remark. The integers k, m in the solution must satisfy $k = 2m - 2$, $m \geq 3$, and $a = 1$, $b = 2^{m-1} - 1$, $c = 2^{m-1} + 1$, $d = 2^{2m-2} - 1$.

Example 8. Two players by turns draw diagonals in a regular $(2n + 1)$-gon ($n > 1$). It is forbidden to draw a diagonal, which was already drawn, or intersect an odd number of already drawn diagonals. The player, who has no legal move, loses. Who has a winning strategy? (Problem of 2007 All-Russian Mathematical Olympiad).

Solution. Let us call the first player A and the second B. The answer is: if n is odd, then B has a winning strategy. If n is even, then A has.

The sum of the number of vertices on two sides of any diagonal of a regular $(2n+1)$-gon is an odd number $2n - 1$. So, there are an even number of vertices on one side of the diagonal. Thus, each diagonal intersects with an even number of diagonals.

If at a time that the game cannot go on, then each undrawn diagonal intersects with an odd number of drawn diagonals and an odd number of undrawn diagonals as well. This situation can only occur when the number of undrawn diagonals is even. (In fact, each undrawn diagonal intersects with odd undrawn diagonals. If the number of undrawn diagonals is odd, then the total number of intersection points of undrawn diagonals is odd. Since each intersection point is counted twice, the total number of intersection points is even, a contradiction.) Therefore, the first player A wins if and only if the total number of diagonals is odd.

There are totally $\frac{(2n+1)(2n-2)}{2} = (n-1)(2n+1)$ diagonals in a $(2n+1)$-gon. Hence, if n is even, A has a winning strategy, and if n is odd, B has.

Exercises

1. Take any permutation $\{a_1, a_2, \ldots, a_{81}\}$ of $\{1, 2, 3, 4, \ldots, 80, 81\}$. Let $\{b_1, b_2, \ldots, b_{27}\}$ be any permutation of the following 27 numbers:
$$|a_1 - a_2 + a_3|, |a_4 - a_5 + a_6|, \ldots, |a_{79} - a_{80} + a_{81}|.$$
Then let $\{c_1, c_2, \ldots, c_9\}$ be any permutation of the following nine numbers:
$$|b_1 - b_2 + b_3|, |b_4 - b_5 + b_6|, \ldots, |b_{25} - b_{26} + b_{27}|.$$
continuing this way, until we get only one number x. Is x odd or even?

2. Let integers $a_1, a_2, \ldots, a_n$ satisfy $a_1 a_2 \ldots a_n = n$ and $a_1 + a_2 + \cdots + a_n = 0$. Prove that $4 \mid n$.

3. In the expression of
$$rvz - rwy - suz + swx + tuy - tvx, \tag{1}$$
$r, s, t, u, v, w, x, y, z$ take values of either 1 or -1.

   (a) Prove that the value of (1) is even.
   (b) Find the maximal possible value of (1).

4. Let A be a 3×3 array consisting of the numbers $1, 2, 3, \ldots, 9$. Compute the sum of the three numbers on the ith row of A and the sum of three numbers on the jth column of A. The number at the intersection of the ith row and jth column of another 3×3 array B is equal to the absolute difference of the two sums of array A. For example,
$$b_{12} = |(a_{11} + a_{12} + a_{13}) - (a_{12} + a_{22} + a_{32})|.$$
Is it possible to arrange the numbers in array A so that the numbers $1, 2, 3, \ldots, 9$ will also appear in array B?

a_{11}	a_{12}	a_{13}
a_{21}	a_{22}	a_{23}
a_{31}	a_{32}	a_{33}

(a)

b_{11}	b_{12}	b_{13}
b_{21}	b_{22}	b_{23}
b_{31}	b_{32}	b_{33}

(b)

(Problem of Team Contest of 2007 Invitational World Youth Mathematics Intercity Competition, Changchun, China).

5. Let n be a positive integer and let $d_1, d_2, \ldots, d_k$ be all its positive divisors, such that $1 = d_1 < d_2 < \cdots < d_k$. Find all values of n for which $k \geq 4$ and $n = d_1^2 + d_2^2 + d_3^2 + d_4^2$ (Problem of 1989 Balkan Mathematical Olympiad).

6. For every sequence $s = (x_1, x_2, \ldots, x_n)$ of natural numbers $\{1, 2, \ldots, n\}$ arranged in any order, denote by $f(s)$ the minimal absolute values of the differences between two consecutive members of s. Find the maximal value of $f(s)$ where s runs through the set of all such sequences (Problem of 1989 International Mathematical Olympiad Longlist).

7. Prove that in the sequence $\{a_n\}$, $a_{n+1} = \left\lfloor \frac{3a_n}{2} \right\rfloor, n \in \mathbb{N}^+$, with $a_1 = 2$, there are infinitely many odd numbers and infinitely many even numbers.

8. Prove: If the sum of all positive divisors of $n \in \mathbb{N}^+$ is a power of two then the number/amount of the divisors is a power of two (Problem of 2009 Middle European Mathematical Olympiad).

9. There are $2n + 1$ bags of balls. If any bag is taken away, we can always divide the remaining $2n$ bags in two groups such that each group has n bags and an equal number of balls in total. Prove that each bag contains the same number of balls.

Chapter 8

Area Methods

Area is an important concept in plane geometry. When dealing with some geometricproblems, the method that takes area as the starting point of calculation or argument is called area method.

The area formula can not only be used to calculate area or to prove the area relation, but also be used to prove that geometry propositions have nothing to do with area (almost all calculation and proofs in plane geometry can be solved by area), sometimes it can get twice the result with half the effort.

The triangle area formula is one of the most fundamental ones, and has many variants. With these formulas, we can not only derive many area formulas of other figures, but also get some property theorems related to area, so as to mutually invert the line segment ratio and the area ratio.

Example 1. In an isosceles right-angled triangle $\triangle ABC$, $\angle BAC = 90°$, point D is the midpoint of side AC, and the line passing through A and perpendicular to BD meets the side BC at point F. Prove that $BF = 2FC$.

Solution. Since $\frac{S_{\triangle ABF}}{S_{\triangle ABD}} = \frac{AB \cdot AF}{AD \cdot BD}$ by $\angle BAF = \angle ADB$, and $\frac{S_{\triangle ABD}}{S_{\triangle AFC}} = \frac{AB \cdot BD}{AF \cdot AC}$ by $\angle FAC = \angle ABD$ (see Fig. 8.1), thus,

$$\frac{BF}{FC} = \frac{S_{\triangle ABF}}{S_{\triangle AFC}} = \frac{S_{\triangle ABF}}{S_{\triangle ABD}} \cdot \frac{S_{\triangle ABD}}{S_{\triangle AFC}} = \frac{AB}{AD} \cdot \frac{AB}{AC} = 2.$$

That is, $BF = 2FC$.

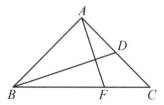

Figure 8.1

Example 2. In a convex quadrilateral $ABCD$, take points E and F on the sides AB and BC, respectively, such that DE, DF divide the diagonal AC into three equal parts. If $S_{\triangle ADE} = S_{\triangle CDF} = \frac{1}{4}S_{ABCD}$, then prove that $ABCD$ is a parallelogram (Problem from the 1990 All-Russian Mathematical Contest for Grade 11).

Solution. Let DE and DF intersect AC at points P and Q, respectively. Let line BD intersect AC at point M.

Since $S_{\triangle ADP} = S_{\triangle CDQ}$ by $AP = QC$ and $S_{\triangle ADE} = S_{\triangle CDF}$, we have $S_{\triangle AEP} = S_{\triangle CFQ}$.

Hence, the distances from points E and F to AC are equal. Therefore, $EF \parallel AC$.

Let $\frac{AB}{AE} = \frac{CB}{CF} = k$, then

$$\frac{S_{\triangle ADB}}{S_{\triangle ADE}} = \frac{AB}{AE} = k, \quad \frac{S_{\triangle CDB}}{S_{\triangle CDF}} = \frac{CB}{CF} = k.$$

Thus,

$$S_{ABCD} = S_{\triangle ADB} + S_{\triangle CDB} = k(S_{\triangle ADE} + S_{\triangle CDF})$$

$$= k \cdot \frac{1}{2}S_{ABCD} \text{ (Fig. 8.2)}.$$

That is, $k = 2$. Hence, $\frac{AQ}{AP} = \frac{AB}{AE} = 2$, therefore, $BQ \parallel EP$. In the same way, we have $BP \parallel FQ$. So $BPDQ$ is a parallelogram, and $BM = MD$,

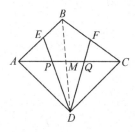

Figure 8.2

$PM = MQ$. Since $AP = QC$, $AM = MC$. That is, AC and BD bisect each other at M. Therefore, $ABCD$ is a parallelogram.

Remark. Some important parallel relations in the proof of Example 2 are derived from the area relation: first, $EF \parallel AC$ is obtained by the area relation, and then the area equation that S_{ABCD} satisfies is obtained by the ratio of the segments, which is used to prove that $BQ \parallel EP$ and $BP \parallel FQ$. The characteristic of area method is to connect all known quantities and unknown quantities with area formulas, so as to change the geometric relations into algebraic relations. Thus, calculation of areas can reduce the difficulty of analyzing problems or adding auxiliary lines, and make the proof simple and concise.

Example 3. In $\triangle ABC$, denote the lengths of sides opposite to vertices A, B, and C by a, b, and c, respectively. The distances from vertices A, B, and C to the incentre I of $\triangle ABC$ are l, m, and n, respectively (Fig. 8.3). Prove that

$$al^2 + bm^2 + cn^2 = abc.$$

Solution. Let the incircle of $\triangle ABC$ touch sides BC, CA, AB at points D, E, and F, respectively.

Since $\angle AFI = \angle AEI = 90°$, the quadrilateral $AEIF$ is inscribed in a circle with diameter AI. Obviously, $AI \perp EF$, by the area formula of the quadrilateral,

$$S_{AEIF} = \frac{1}{2} AI \cdot EF = \frac{1}{2} AI \cdot AI \sin A$$

$$= \frac{1}{2} l^2 \cdot \frac{a}{2R} = \frac{al^2}{4R}.$$

where the second and the third equalities are from the law of sine applied to $\triangle AEF$ and $\triangle ABC$, respectively, and R is the circumradius of $\triangle ABC$.

Similarly, we have $S_{BFID} = \frac{bm^2}{4R}$, $S_{CDIE} = \frac{cn^2}{4R}$.

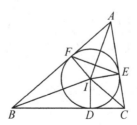

Figure 8.3

Thus, $S_{\triangle ABC} = S_{AEIF} + S_{BFID} + S_{CDIE} = \frac{al^2 + bm^2 + cn^2}{4R}$.

On the other hand, $S_{\triangle ABC} = \frac{abc}{4R}$, therefore, $al^2 + bm^2 + cn^2 = abc$.

Remark. The reader might as well prove that $\frac{S_{\triangle ABC}}{S_{\triangle DEF}} = \frac{2R}{r}$, where R and r are the circumradius and the inradius of incircle of $\triangle ABC$, respectively.

A basic idea of the area method is that two different methods of calculating the same area should yield the same result. In the above problem, divide the triangle into three quadrilaterals and establish the area equation. Considering that these three quadrilaterals all have circumcircles, and the diagonals are perpendicular to each other, there is no real difficulty in expressing their areas with known quantities, and the introduced circumradius R can eliminate the sine of the angle and plays a transitional role.

The area formula of atriangle or quadrilateral reveals the inner relations between basic elements such as edge and angle, and the area of a triangle is often related to the incentre and circumcentre.

Example 4. In an inscribed hexagon $ABCDEF$, $AB \cdot CD \cdot EF = BC \cdot DE \cdot FA$. Show that three lines AD, BE, CF are concurrent.

Solution. Draw lines AC, CE, and EA. Let AC, CE, and EA meet BE, AD, and CF at points P, Q, and R, respectively.

In the inscribed quadrilateral $ABCE$, since $\angle BAE$ and $\angle BCE$ are supplementary, $\sin \angle BAE = \sin \angle BCE$. Therefore, by the common edge theorem and the triangle area formula, we obtain (Fig. 8.4)

$$\frac{AP}{PC} = \frac{S_{\triangle BAE}}{S_{\triangle BCE}} = \frac{AB \cdot AE}{BC \cdot CE}.$$

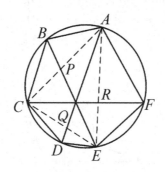

Figure 8.4

In the same way, we have

$$\frac{CQ}{QE} = \frac{S_{\triangle CAD}}{S_{\triangle EAD}} = \frac{AC \cdot CD}{AE \cdot DE},$$

$$\frac{ER}{RA} = \frac{S_{\triangle ECF}}{S_{\triangle ACF}} = \frac{CE \cdot EF}{AC \cdot FA}.$$

Multiply both sides of the above three equalities, to get

$$\frac{AP}{PC} \cdot \frac{CQ}{QE} \cdot \frac{ER}{RA} = \frac{AB \cdot AE}{BC \cdot CE} \cdot \frac{AC \cdot CD}{AE \cdot DE} \cdot \frac{CE \cdot EF}{AC \cdot FA} = \frac{AB \cdot CD \cdot EF}{BC \cdot DE \cdot FA} = 1.$$

Since lines AD, BE, and CF are not parallel, by the converse Ceva theorem, three lines AD, BE, CF are concurrent.

Remark. Ceva's theorem has several equivalent forms. If using 'the angular form of Ceva's theorem', the proof will be more concise.

Example 5. The extensions of the edges AB and DC of a convex quadrilateral $ABCD$ meet at point E, the extensions of edges AD and BC meet at point F. Prove that the midpoints M, N, and L of AC, BD, and EF, respectively, are collinear. (The line MNL is called *the Newton line* of the complete quadrilateral $ABCDEF$.)

Solution. Draw the segments MB, MD, ME, MF, NE, NF, and MN.
Since M and N are midpoints of AC and BD, respectively (Fig. 8.5).

$$S_{\triangle MDE} = S_{\triangle MNE} + S_{\triangle MDN} + S_{\triangle EDN}$$

$$= S_{\triangle MNE} + S_{\triangle MBN} + S_{\triangle EBN} = 2S_{\triangle MNE} + S_{\triangle MBE}.$$

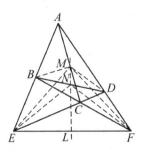

Figure 8.5

So

$$S_{\triangle MNE} = \frac{S_{\triangle MDE} - S_{\triangle MBE}}{2}$$

$$= \frac{S_{\triangle ADE} - S_{\triangle CBE}}{4} = \frac{1}{4}S_{ABCD}.$$

In the same way,

$$S_{\triangle MNF} = \frac{S_{\triangle MBF} - S_{\triangle MDF}}{2} = \frac{S_{\triangle ABF} - S_{\triangle CDF}}{4} = \frac{1}{4}S_{ABCD}.$$

Therefore, $S_{\triangle MNE} = S_{\triangle MNF}$.

Since points E and F are on different sides of the line MN (under the known condition of the figure, line MN intersects with segment EF), therefore line MN bisects segment EF, that is, M, N, and L are collinear.

Remark. In this problem, we first make full use of the condition that 'M and N are midpoints of AC and BD, respectively.' to conduct area transformations, and finally reach another midpoint L on line MN by using $S_{\triangle MNE} = S_{\triangle MNF}$, which in fact is based on the following theorem:

Theorem. *Let P be a point on the plane of $\triangle ABC$. Line CP intersects with AB (or its extension) at point D. Then $\frac{S_{\triangle APC}}{S_{\triangle BPC}} = \frac{AD}{BD}$ (Fig. 8.6).*

So, conversely if P is inside $\triangle ABC$, D is on AB and $\frac{S_{\triangle APC}}{S_{\triangle BPC}} = \frac{AD}{BD}$, then C, P, and D are collinear. By this conclusion, the last step of the proof of Example 5 follows without difficulty. So, the area method is one of the ways to prove collinearity of three points.

In combinatorial geometry, we often consider some area problems. The following example is a covering problem, in which the principle of area overlapping *is used in the proof:*

Put n regions of area S_i $(1 \leq i \leq n)$ into a region C of area S_0. If the sum of the n areas is greater than kS_0, then there is a point in C covered by at least $k + 1$ regions. If the sum of the n areas is less than kS_0, then there is a point in C covered by at most $k - 1$ regions.

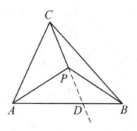

Figure 8.6

Example 6. There are 650 red points in a circle with radius 16. Prove that you can make a ring C with inner radius 2 and outer radius 3 such that at least 10 red points are in the ring C (the boundary excluded).

Solution. Make ring C_i $(1 \le i \le 650)$ of inner radius $2+\varepsilon$ and outer radius $3-\varepsilon$ with centre at each red point. The area of C_i is

$$S_i = \pi \left((3-\varepsilon)^2 - (2+\varepsilon)^2 \right) = 5\pi(1-2\varepsilon),$$

where $0 < \varepsilon < 1/2$.

Obviously, these rings are within a circle C_0 with radius 19, and the area of C_0 is $S_0 = 361\pi$.

Take $\varepsilon = 0.0001$, then,

$$\sum_{i=1}^{650} S_i = 3250\,\pi \times 0.9998 > 3249\,\pi = 9S_0.$$

So, by the *principle of area overlapping*, there must be a point in C_0 covered by at least 10 rings C_i. Make a ring C of inner radius 2 and outer radius 3 centred at this point, and C covers at least 10 red points.

Remark. The *principle of area overlapping* can be treated as one of the drawer principle in geometry. When considering some problems of covering, embedding, and overlapping, we often prove the existence by area (sometimes supplemented by the technique of expansion and contraction).

Example 7. On a plane rectangular coordinate system, there are n $(n \ge 3)$ lattice points lying on a circle O of radius R. Prove that $n < 2\pi \cdot \sqrt[3]{R^2}$.

Solution. Arrange all lattice points anticlockwise on the circle by $A_1, A_2, \ldots, A_n$ $(n \ge 3)$. Let $A_{n+1} = A_1$, $A_{n+2} = A_2$. For $i = 1, 2, \ldots, n$, since $\triangle A_i A_{i+1} A_{i+2}$ is a lattice triangle, the area of which $S_{\triangle A_i A_{i+1} A_{i+2}} \ge 1/2$.

On the other hand, let $\angle A_i O A_{i+1} = \theta_i$, $i = 1, 2, \ldots, n$, then

$$|A_i A_{i+1}| = 2R\sin\frac{\theta_i}{2}, \quad |A_{i+1} A_{i+2}| = 2R\sin\frac{\theta_{i+1}}{2},$$

$$\angle A_i A_{i+1} A_{i+2} = \frac{\pi - \theta_i}{2} + \frac{\pi - \theta_{i+1}}{2} = \pi - \frac{\theta_i + \theta_{i+1}}{2}.$$

Thus,

$$S_{\triangle A_i A_{i+1} A_{i+2}} = \frac{1}{2} \cdot |A_i A_{i+1}| \cdot |A_{i+1} A_{i+2}| \cdot \sin \angle A_i A_{i+1} A_{i+2}$$

$$= \frac{1}{2} \cdot 2R \sin \frac{\theta_i}{2} \cdot 2R \sin \frac{\theta_{i+1}}{2} \cdot \sin \frac{\theta_i + \theta_{i+1}}{2}$$

$$< 2R^2 \cdot \frac{\theta_i}{2} \cdot \frac{\theta_{i+1}}{2} \cdot \frac{\theta_i + \theta_{i+1}}{2}.$$

Therefore,

$$\frac{1}{2} < \frac{R^2}{4} \cdot \theta_i \theta_{i+1} (\theta_i + \theta_{i+1}) \leq \frac{R^2}{16} (\theta_i + \theta_{i+1})^3.$$

Consequently,

$$\theta_i + \theta_{i+1} > \frac{2}{\sqrt[3]{R^2}}. \tag{1}$$

Note that $\sum_{i=1}^{n} \theta_i = 2\pi$, summing up (1) for $i = 1, 2, \ldots, n$, we have $4\pi > n \cdot \frac{2}{\sqrt[3]{R^2}}$, that is, $n < 2\pi \cdot \sqrt[3]{R^2}$.

Remark. The reader might consider how to extend the result to the case for an ellipse.

Exercises

1. Consider triangle $P_1 P_2 P_3$ and a point P within the triangle. Lines $P_1 P, P_2 P, P_3 P$ intersect the opposite sides in points Q_1, Q_2, Q_3, respectively. Prove that, of the numbers $\frac{P_1 P}{PQ_1}, \frac{P_2 P}{PQ_2}, \frac{P_3 P}{PQ_3}$, at least one is ≤ 2 and at least one is ≥ 2 (Problem from the 1961 International Mathematical Olympiad).

2. Prove that any line that bisects the area and the perimeter of a triangle simultaneously must pass through the incentre of the triangle.

3. In quadrilateral $ABCD$, the area ratios of $\triangle ABD$, $\triangle BCD$, $\triangle ABC$ are 3:4:1. Points M and N are on AC and CD, respectively, such that $\frac{AM}{AC} = \frac{CN}{CD}$ and points B, M, and N are collinear. Prove that M and N are midpoints of AC and CD, respectively (see Fig. 8.7) (Problem from the 1983 China High School Mathematical League, Test 2).

4. Let line l pass through the centroid G of $\triangle ABC$, and meet the sides AB and AC at points B_1 and C_1, respectively, such that $\frac{AB_1}{AB} = \lambda$, $\frac{AC_1}{AC} = \mu$. Prove that $\frac{1}{\lambda} + \frac{1}{\mu} = 3$.

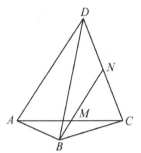

Figure 8.7

5. In a convex hexagon $ABCDEF$, points P, Q, R, S, T, and U are midpoints of edges AB, BC, CD, DE, EF, and FA, respectively. If PS, QT, RU all bisect the area of the hexagon $ABCDEF$, prove that PS, QT, RU are concurrent.

6. Denote the area of a convex quadrilateral $ABCD$ by S. The quadrilateral $A'B'C'D'$ is obtained such that A' and A, C' and C are symmetric about the diagonal BD; B' and B, D' and D are symmetric about the diagonal AC. Denote the area of quadrilateral $A'B'C'D'$ by S'. Prove that $S'/S < 3$ (Problem from the 2005 All-Russian Mathematical Olympiad, Round 4).

7. In $\triangle ABC$, M is the midpoint of BC. Three points N, P, and Q are on segments AM, AB, and AC (not endpoints), respectively, such that A, P, N, Q are concyclic (see Fig. 8.8). Prove that $AP \cdot AB + AQ \cdot AC = 2AN \cdot AM$.

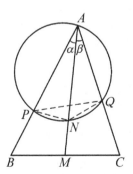

Figure 8.8

8. In a convex pentagon $ABCDE$, denote by $AD \cap BE = F$, $BE \cap CA = G$, $CA \cap DB = H$, $DB \cap EC = I$, $EC \cap AD = J$, $AI \cap BE = A'$, $BJ \cap CA = B'$, $CF \cap DB = C'$, $DG \cap EC = D'$, $EH \cap AD = E'$. Prove that $\frac{AB'}{B'C} \cdot \frac{CD'}{D'E} \cdot \frac{EA'}{A'B} \cdot \frac{BC'}{C'D} \cdot \frac{DE'}{E'A} = 1$ (Problem from the 2009 China Team Selection Test).

Chapter 9

Thinking Globally

In solving some mathematical problems, we should think globally of the structure and essential character of the problem.

The main strategies for globally thinking for solving problems are:

First, if the condition or conclusion of the problem possesses a global character, we should use its essential property so as to simplify the problem.

Second, if the condition or conclusion of a problem is local, then change the problem into a global one.

Let us get a feel for the strategy by two simple examples.

Example 1. Two sequences $\{a_n\}$ and $\{b_n\}$ with $a_1 = 1$, $b_1 = 3$ satisfy

$$\begin{cases} a_{n+1} = a_n + b_n - \sqrt{a_n^2 - a_n b_n + b_n^2}, \\ b_{n+1} = b_n + a_n + \sqrt{b_n^2 - b_n a_n + a_n^2}. \end{cases}$$

Find the general terms of $\{a_n\}$ and $\{b_n\}$.

Solution. By the definition of the sequences,

$$a_{n+1} + b_{n+1} = 2(a_n + b_n).$$

$$a_{n+1} b_{n+1} = (a_n + b_n)^2 - (a_n^2 - a_n b_n + b_n^2) = 3a_n b_n.$$

Thus, $a_n + b_n = 2^{n-1}(a_1 + b_1) = 2^{n+1}$ and $a_n b_n = 3^{n-1} a_1 b_1 = 3^n$.

So, $a_n < b_n$ are the two real roots of the equation $x^2 - 2^{n+1}x + 3^n = 0$. Therefore,

$$a_n = 2^n - \sqrt{4^n - 3^n}, \quad b_n = 2^n + \sqrt{4^n - 3^n}.$$

Example 2. Suppose that there are nonzero real numbers a, b, c such that the two triple sets $A = \{ab, 2b, 3c\}$, $B = \{a, 2b^2, 3bc\}$ are equal. Find all possible values of b.

Solution 1. Let $A = B$. There are three cases for $a \in B$.

If $a = ab$, then $b = 1$.

If $a = 2b$, then $2b^2 = ab$, so $3bc = 3c$, we also have $b = 1$.

If $a = 3c$, then $3bc = ab$, so $2b^2 = 2b$, we also have $b = 1$.

To Check: take $b = 1$, then $A = \{a, 2, 3c\} = B$ as long as $a \neq 2$, $c \neq 2/3$, and $a \neq 3c$. Therefore, $b = 1$.

Solution 2. Let $A = B$, then the two products of all elements of each set are equal, that is $ab \cdot 2b \cdot 3c = a \cdot 2b^2 \cdot 3bc$, so $b = 1$.

The checking part is the same as for Solution 1.

Remark. Solution 2 used a global strategy and is simpler than Solution 1.

Example 3. Let $M = \{1, 2, \ldots, 19\}$ and $A = \{a_1, a_2, \ldots, a_k\} \subset M$. Find the least k so that for any $b \in M$, there exist $a_i, a_j \in A$, satisfying $a_i = b$ or $a_i \pm a_j = b$ (a_i and a_j do not have to be different) (Problem from the 2006 China Girls' Mathematical Olympiad).

Solution. Since there are $k + k + 2 \dbinom{k}{2} = k(k+1)$ ways of taking one or two elements from set A, we see that $k(k+1) \geq 19$, that is $k \geq 4$.

If $k = 4$, suppose that $A = \{a_1, a_2, a_3, a_4\}$ with $a_1 < a_2 < a_3 < a_4$ satisfying the condition of the problem. Since among the 20 numbers of $\{a_i | i = 1, 2, 3, 4\}$, $\{2a_i | i = 1, 2, 3, 4\}$, and $\{a_i \pm a_j | 1 \leq j < i \leq 4, i, j \in \mathbb{N}^+\}$, $2a_4$ is the largest, so $2a_4 \geq 19$, it follows that $2a_4 \geq 20$ since a_4 is an integer.

Thus, 19 numbers of $a_i (i = 1, 2, 3, 4)$, $2a_i (i = 1, 2, 3)$, $a_i \pm a_j (1 \leq j < i \leq 4)$ are all 19 members of M. Simplifying the summation of the 19 numbers, we have

$$3a_1 + 5a_2 + 7a_3 + 7a_4 = 1 + 2 + \cdots + 19 = 190. \tag{1}$$

The largest number in these 19 numbers is obviously

$$a_3 + a_4 = 19, \tag{2}$$

Hence, $3a_1 + 5a_2 = 190 - 7 \times 19 = 57$, thus $3 \,|\, a_2$. On the other hand, by (2), we have $a_2 < a_3 \leq 9$, thus $a_2 \leq 6$, but then $3a_1 + 5a_2 < 8a_2 \leq 48 < 57$, a contradiction!

Therefore, $k_{\min} \geq 5$.

It is easy to verify that $A = \{1, 3, 5, 9, 16\}$ satisfies the conditions of the problem, so $k_{\min} = 5$.

Remark. In denying $k = 4$, equation (1) is obtained from global thinking, which makes it simpler than discussing the eight possible values of a_4.

Example 4. To each vertex of a regular pentagon, an integer is assigned in such a way that the sum of all five numbers is positive. If three consecutive vertices are assigned the numbers x, y, z, respectively and $y < 0$, then the following operation is allowed: the numbers x, y, z are replaced by $x+y$,-y, $z+y$, respectively. Such an operation is performed repeatedly as long as at least one of the five numbers is negative. Determine whether this process necessarily comes to an end after a finite number of steps (Problem from the 1986 International Mathematical Olympiad).

Solution. The answer is positive. We see that the sum of the five numbers is unchanged after the operation. If all numbers are non-negative, then no operation can be performed.

If there is a $y < 0$, then we may write the circular five numbers in a line for convenience: v, w, x, y, z (where z and v are adjacent). After an operation, the numbers change to v, w, $x+y$, $-y$, $z+y$. Consider a global symmetric quantity:

$$f(u, v, w, x, y, z) = v^2 + w^2 + x^2 + y^2 + z^2 + (v + w)^2 + (w + x)^2$$
$$+ (x + y)^2 + (y + z)^2 + (z + v)^2.$$

The difference of the two quantities before and after the operation is

$$f(v, w, x + y, -y, z + y) - f(v, w, x, y, z) = 2y(v + w + x + y + z) < 0.$$

Thus, the positive integer quantity is strictly decreasing under each operation, but it cannot decrease forever. Therefore, the procedure necessarily comes to an end, so that all five numbers are non-negative.

Remark. There is another global symmetric structure:

let the consecutive five numbers be u_1, u_2, u_3, u_4, u_5, where u_1, u_5 are adjacent.

$$f(u_1, u_2, u_3, u_4, u_5) = \sum_{i=1}^{5} |u_i| + \sum_{i=1}^{5} |u_i + u_{i+1}| + \sum_{i=1}^{5} |u_i + u_{i+1} + u_{i+2}|$$

$$+ \sum_{i=1}^{5} |u_i + u_{i+1} + u_{i+2} + u_{i+3}|,$$

where $u_{5+i} = u_i, i = 1, 2, 3$. We can prove that f is strictly decreasing under the operation.

Example 5. On an $n \times n$ $(n \geq 4)$ chart, place '+' signs in the cells of the main diagonal and '−' signs in all the other cells. You can change all the signs in one row or in one column, from '−' to '+' or '+' to '−'. Prove that you shall always have n or more '+' signs after finitely many operations (Problem from the 2010 All-Russian Mathematical Olympiad).

Solution. We use (i, j) to indicate the cell in the ith row and the jth column.

At the beginning, cell (i, i) has the sign of '+', $i = 1, 2, \ldots, n$.

Consider n disjoint groups of cells: $\{(a, c), (a, d), (b, c), (b, d)\}$, where $a = c = i, b \equiv i + 1 \pmod{n}, d \equiv i + 2 \pmod{n}$. $i = 1, 2, \ldots, n$.

In each group, there is just one cell with sign '+' at the beginning. After any number of operations, we see that each group has at least one sign of '+'. Therefore, there are at least n signs of '+' in the chart after any number of operations.

Example 6. Prove that for any n real numbers $r_1, r_2, \ldots, r_n$, there is a subset S of $\{1, 2, \ldots, n\}$, satisfying $1 \leq |S \cap \{i, i+1, i+2\}| \leq 2$ for any $i \in \{1, 2, \ldots, n-2\}$ and $|\sum_{i \in S} r_i| \geq \frac{1}{6} \sum_{i=1}^{n} |r_i|$.

Solution. Denote $s = \sum_{i=1}^{n} |r_i|$. Let $s_i = \sum_{\substack{r_j \geq 0, \\ j \equiv i \pmod 3}} r_j$, $t_i = \sum_{\substack{r_j < 0, \\ j \equiv i \pmod 3}} r_j$, where $i = 1, 2, 3$. Then $s = s_1 + s_2 + s_3 - t_1 - t_2 - t_3$, so we have

$$2s = (s_1 + s_2) + (s_2 + s_3) + (s_3 + s_1) - (t_1 + t_2) - (t_2 + t_3) - (t_3 + t_1).$$

Thus, there exist $a, b \in \{1, 2, 3\}$, $a \neq b$, such that $s_a + s_b \geq \frac{s}{3}$ or $t_a + t_b \leq -\frac{s}{3}$ holds. Without loss of generality, suppose that $|s_a + s_b| \geq |t_a + t_b|$ (otherwise interchange s_i and t_i, $i = 1, 2, \ldots, n$), so $s_a + s_b \geq \frac{s}{3}$, then,

$$|s_a + s_b + t_a| + |s_a + s_b + t_b| = 2(s_a + s_b) + (t_a + t_b) \geq \frac{s}{3}.$$

If $|s_a + s_b + t_a| \geq |s_a + s_b + t_b|$, then $|s_a + s_b + t_a| \geq \frac{s}{6}$, take

$$S = \{j | j \equiv a \pmod 3\} \cup \{j | r_j \geq 0, \ j \equiv b \pmod 3\};$$

if $|s_a + s_b + t_a| < |s_a + s_b + t_b|$, then $|s_a + s_b + t_b| \geq \frac{s}{6}$, take

$$S = \{j | r_j \geq 0, \ j \equiv a \pmod 3\} \cup \{j | j \equiv b \pmod 3\}.$$

In either case, we have, $1 \leq |S \cap \{i, i+1, i+2\}| \leq 2$ for any $i \in \{1, 2, \ldots, n-2\}$, and $\left|\sum_{i \in S} r_i\right| \geq \frac{s}{6} = \frac{1}{6} \sum_{i=1}^{n} |r_i|$.

Exercises

1. Given an array

$$\begin{pmatrix} -1 & 2 & -3 & 4 \\ -1.2 & 0.5 & -3.9 & 9 \\ \pi & -12 & 4 & -2.5 \\ 63 & 1.4 & 7 & -9 \end{pmatrix}$$

A transformation of the array is to change signs of all numbers of a row or a column of the array. Can all the numbers in the array be positive after some transformations?

2. Let $a_1, a_2, \ldots, a_9$ be nonzero real numbers. Prove that there is at least one number that is negative among the following six numbers:

$$a_1 a_5 a_9, \quad a_2 a_6 a_7, \quad a_3 a_4 a_8, \quad -a_3 a_5 a_7, \quad -a_1 a_6 a_8, \quad -a_2 a_4 a_9.$$

3. Let $S = \{1, 2, \ldots, 21\}$. Suppose that there is a finite set $A \subset \mathbb{N}^+$, such that any element of S is either a member of A or is equal to the sum of two different elements of A. Find the minimal number of elements of A.

4. Suppose that real numbers x_1, x_2, x_3, x_4, x_5 satisfy the system

$$\begin{cases} x_1 x_2 + x_1 x_3 + x_1 x_4 + x_1 x_5 = -1, \\ x_2 x_1 + x_2 x_3 + x_2 x_4 + x_2 x_5 = -1, \\ x_3 x_1 + x_3 x_2 + x_3 x_4 + x_3 x_5 = -1, \\ x_4 x_1 + x_4 x_2 + x_4 x_3 + x_4 x_5 = -1, \\ x_5 x_1 + x_5 x_2 + x_5 x_3 + x_5 x_4 = -1. \end{cases}$$

Find all possible values of x_1 (Problem from the 2009 Japan Mathematical Olympiad Preliminary).

5. Given $n\,(n > 1)$ quadratic polynomials $x^2 - a_i x + b_i\,(1 \leq i \leq n)$, where $2n$ real numbers a_i, b_i are distinct, is it possible that each number of a_i, $b_i\,(1 \leq i \leq n)$ is a root of some polynomial in the n polynomials? (Problem from the fourth round of 2006 All-Russian Mathematical Olympiad).

6. There are $n\,(\geq 4)$ positive integers lying on a circle. If there are four consecutive numbers a, b, c, d on the circle satisfying $(a - d)(b - c) > 0$, then swapping b and c is called an *operation*.

   Determine whether the process necessarily comes to an end after a finite number of *operations*. Explain your reasoning.

7. In an $n \times n (n \geq 2)$ array of real numbers, let the sum of the n numbers in the ith row be s_i, and the sum of the n numbers in the jth column be t_j. Denote $a_{ij} = s_i - t_j$, $i, j = 1, 2, \ldots, n$. Find the maximal possible number of positive real numbers of n^2 numbers a_{ij} $(i, j = 1, 2, \ldots, n)$.

Chapter 10

Proper Representations

A suitable representation is always the key to solving problems.

For example, there are many ways to represent an integer, such as prime factorization, classifying by modulo, number system of some base.

In solving problems, we can choose the proper representation of numbers according to the characteristics of the problem, so as to make the idea clear, or to make further study of the problem easier.

Example 1. Positive integers p, q, r, a satisfy $pq = ra^2$, where r is prime, p and q are coprime. Show that one of p and q is a perfect square.

Solution. Let $p = p_1^{\alpha_1} p_2^{\alpha_2} \ldots p_k^{\alpha_k}$, $q = q_1^{\beta_1} q_2^{\beta_2} \ldots q_l^{\beta_l}$, $a = a_1^{\gamma_1} a_2^{\gamma_2} \ldots a_m^{\gamma_m}$ be the prime factorizations of p, q, a, respectively. Then,

$$p_1^{\alpha_1} \cdot p_2^{\alpha_2} \ldots p_k^{\alpha_k} \cdot q_1^{\beta_1} \cdot q_2^{\beta_2} \ldots q_l^{\beta_l} = ra_1^{2\gamma_1} a_2^{2\gamma_2} \ldots a_m^{2\gamma_m}.$$

Since p and q are coprime and r is prime, one of p, q that is not divisible by r is a perfect square, since each of its prime factors has an even power.

Remark. The fundamental theorem of arithmetic is very important, which says that any integer $n > 1$ has a unique prime factorization

$$n = p_1^{\alpha_1} p_2^{\alpha_2} \ldots p_k^{\alpha_k},$$

where $p_1 < p_2 < \cdots < p_k$ are distinct primes, $\alpha_1, \alpha_2, \ldots, \alpha_k \in \mathbb{N}^+$ (or $\mathbb{N}$ for convenience).

We list some consequences of the theorem as follows:

(I) The number of positive factors of n is $d(n) = \prod_{i=1}^{k} (\alpha_i + 1)$, specially, n is a square number if and only if $d(n)$ is odd.

(II) The sum of all positive factors of n is $\sigma(n) = \prod\limits_{i=1}^{k} \frac{p_i^{\alpha_i+1}-1}{p_i-1}$.

(III) The number of positive integers that $<n$ and are coprime to $n \geq 2$ is

$$\varphi(n) = \prod_{i=1}^{k} p_i^{\alpha_i-1}(p_i - 1).$$

Example 2. Let n be a positive integer. Prove that:

(1) If $2n + 1$ is a perfect square, then $n + 1$ is the sum of two consecutive squares.
(2) If $3n + 1$ is a perfect square, then $n + 1$ is the sum of three perfect squares.

Solution. (1) Since $2n+1 = a^2$ is odd, a is odd. So, let $a = 2k+1$, where k is an integer. Then $2n + 1 = (2k + 1)^2$, we have $n + 1 = 2k^2 + 2k + 1 = k^2 + (k + 1)^2$.

(2) Let $3n+1 = a^2$, where a is an integer not divisible by 3. Thus $a = 3k\pm1$, where k is an integer. That is, $3n+1 = (3k\pm1)^2$. So, $n+1 = 3k^2\pm2k+1 = k^2 + k^2 + (k \pm 1)^2$.

Example 3. Is there a 2000-element subset $A \subset S = \{1, 2, \ldots, 3000\}$ such that if $x \in A$ then $2x \notin A$?

Solution. The answer is negative.

Each positive integer can be written in the form of product of an odd number t and a non-negative integer power of 2 as $2^s t$.

If the answer is positive, then, if $2^s t \in A$, then $2^{s+1}t \notin A$.

Thus, for each odd number t, we have

$$\left|A \cap \{t, 2t, 2^2t, \ldots\}\right| \leq \left|S \cap \{t, 2^2t, 2^4t, \ldots\}\right|.$$

Thus, $|A| \leq$ the number of elements of the set:

$$\{1, 3, \ldots, 2999, 1 \times 2^2, 3 \times 2^2, \ldots, 749 \times 2^2, 1 \times 2^4, 3 \times 2^4, \ldots, 187 \times 2^4,$$

$$1 \times 2^6, 3 \times 2^6, \ldots, 45 \times 2^6, 1 \times 2^8, 3 \times 2^8, \ldots, 11 \times 2^8, 1 \times 2^{10}\}.$$

That is,

$$|A| \leq 1500 + 375 + 94 + 23 + 6 + 1 = 1999 < 2000.$$

A contradiction.

Example 4. For each positive integer m, define $f(m) =$ the number of factors 2 in $m!$ (namely, the greatest integer k, such that $2^k \,|\, m!$). Prove that there are infinitely many positive integers m, such that

$$m - f(m) = 1000.$$

Solution. Write m in the form of the sum of powers of 2.

$$m = \sum 2^{r_i} = 2^{r_n} + 2^{r_{n-1}} + \cdots + 2^{r_1},$$

where $r_n > r_{n-1} > \cdots > r_1 \geq 0, r_i \in \mathbb{N}$.

Thus,

$$f(m) = \left\lfloor \frac{m}{2} \right\rfloor + \left\lfloor \frac{m}{2^2} \right\rfloor + \left\lfloor \frac{m}{2^3} \right\rfloor + \cdots$$

$$= \left\lfloor \frac{\sum 2^{r_i}}{2} \right\rfloor + \left\lfloor \frac{\sum 2^{r_i}}{2^2} \right\rfloor + \left\lfloor \frac{\sum 2^{r_i}}{2^3} \right\rfloor + \cdots$$

$$= \sum 2^{r_i - 1} + \sum 2^{r_i - 2} + \sum 2^{r_i - 3} + \cdots,$$

where the sum is taken only for non-negative powers. Further, we have

$$f(m) = \sum (2^{r_i - 1} + 2^{r_i - 2} + \cdots + 1) = \sum (2^{r_i} - 1) = m - n.$$

Thus, $m - f(m) = n$ (the number of 1s of m in binary form.)

Since there are infinitely many positive integers m which have 1000 1s in their binary form, this completes the proof.

Remark. There is an extended conclusion:

For positive integer m, define $f_p(m) =$ the number of prime factor ps in $m!$ (namely the greatest positive integer k such that $p^k \,|\, m!$), then the sum of digits of m in the form of p-decimal number is equal to $\frac{m - f_p(m)}{p - 1}$.

Example 5. Suppose that function $f(n) : \mathbb{N}^+ \to \mathbb{N}^+$ with $f(1) = 1$ satisfies that for any $n \in \mathbb{N}^+$, $\varepsilon \in \{0, 1\}$, $f(2n + \varepsilon) = 3f(n) + \varepsilon$. Find the range of f.

Solution. First find values of $f(n)$ for $n = 1, 2, 3, 4, 5, 6$.

n	1	2	3	4	5	6	...
$f(n)$	1	3	4	9	10	12	...

In the above form, turn n to binary numbers and $f(n)$ to ternary numbers:

n	$(1)_2$	$(10)_2$	$(11)_2$	$(100)_2$	$(101)_2$	$(110)_2$	...
$f(n)$	$(1)_3$	$(10)_3$	$(11)_3$	$(100)_3$	$(101)_3$	$(110)_3$	...

So, we guess that for any positive integer $n = (\overline{a_k a_{k-1} \ldots a_1})_2$, $f(n) = (\overline{a_k a_{k-1} \ldots a_1})_3$.

We prove the number of digits of n by induction on k.

If $k = 1$, the conjecture is true.

If the conjecture is true for $k \geq 1$, then for a $k + 1$ digits number

$$n_1 = (\overline{a_k a_{k-1} \ldots a_0})_2.$$

In $f(2n + \varepsilon) = 3f(n) + \varepsilon$, let $n = (\overline{a_k a_{k-1} \ldots a_1})_2, \varepsilon = a_0 \in \{0, 1\}$, then

$$2n + \varepsilon = 2 \cdot (\overline{a_k a_{k-1} \ldots a_1})_2 + a_0 = (\overline{a_k a_{k-1} \ldots a_1 a_0})_2 = n_1.$$

Thus,

$$f(n_1) = 3f(n) + a_0 = 3 \cdot (\overline{a_k a_{k-1} \ldots a_1})_3 + a_0 = (\overline{a_k a_{k-1} \ldots a_1 a_0})_3.$$

This completes the induction.

Therefore, the range of f_s is the set of all positive integers containing only the digits 0,1 in the ternary representation. That is,

$$\{3^{r_1} + 3^{r_2} + \cdots + 3^{r_s} \, | \, s \in \mathbb{N}^+, r_1, r_2, \ldots, r_s \in \mathbb{N}, r_1 > r_2 > \cdots > r_s\}.$$

Remark. The method above can also be used to solve the following problem from the 1995 China Mathematical Olympiad.

Let $f : \mathbb{N}^+ \to \mathbb{N}^+$ be a function with $f(1) = 1$ such that for any $n \in \mathbb{N}^+$, $3f(n)f(2n + 1) = f(2n)(1 + 3f(n))$ and $f(2n) < 6f(n)$.

Find all solutions of the equation $f(k) + f(l) = 293$, where $k < l$.

We introduce the representation of every positive integer as the sum of elements in the Fibonacci sequence $\{F_n\}$, which is defined by

$$F_1 = 1, \quad F_2 = 2, \quad F_{n+2} = F_{n+1} + F_n, \quad n \in \mathbb{N}^+.$$

For example,

$$10 = 8 + 2 = F_5 + F_2,$$

$$30 = 13 + 8 + 5 + 3 + 1 = F_6 + F_5 + F_4 + F_3 + F_1.$$

But also, we have $30 = 21 + 8 + 1 = F_7 + F_5 + F_1$.

Generally, we can prove the following important theorem by induction.

Theorem 1. *Each positive integer n can be uniquely represented by a sum of some non-adjacent numbers in the Fibonacci sequence $\{F_n\}$, that is, $n = a_k F_k + a_{k-1} F_{k-1} + \cdots + a_1 F_1$ (where $a_1, a_2, \ldots, a_k \in \{0, 1\}$, $a_i a_{i+1} = 0$, $i = 1, 2, \ldots, k - 1$). Usually, we use the notation $n = (\overline{a_k a_{k-1} \ldots a_1})_F$, which is called the Fibonacci representation of n.*

Example 6. Find the number of subsets $A \subset S = \{1, 2, \ldots, n\}$ that contain no consecutive integers.

Solution. By Theorem 1, each set A corresponds to a positive real number $m = (\overline{a_n a_{n-1} \cdots a_1})_F$, where

$$a_i = \begin{cases} 1, & i \in A, \\ 0, & i \notin A. \end{cases}$$

Hence, the number of such subsets A is

$$F_{n+1} - 1 = (\overline{1 \underbrace{00 \cdots 0}_{n \text{ of } 0s}})_F - 1 = \frac{1}{\sqrt{5}} \left(\left(\frac{1 + \sqrt{5}}{2} \right)^{n+2} - \left(\frac{1 - \sqrt{5}}{2} \right)^{n+2} \right) - 1.$$

Example 7. For any given quadrilateral $A_1 A_2 A_3 A_4$, let M, N be the midpoints of diagonals $A_1 A_3$, $A_2 A_4$, respectively. Prove that

$$|A_1 A_2|^2 + |A_2 A_3|^2 + |A_3 A_4|^2 + |A_4 A_1|^2 = |A_1 A_3|^2 + |A_2 A_4|^2 + 4|MN|^2.$$

Solution. Let the Cartesian coordinates of the vertices be $A_i(x_i, y_i)$, $i = 1, 2, 3, 4$, then the coordinates of M, N are $\left(\frac{x_1 + x_3}{2}, \frac{y_1 + y_3}{2} \right)$, $\left(\frac{x_2 + x_4}{2}, \frac{y_2 + y_4}{2} \right)$, respectively.

Then

$$4 \left(\frac{x_1 + x_3}{2} - \frac{x_2 + x_4}{2} \right)^2 + (x_1 - x_3)^2 + (x_2 - x_4)^2$$

$$= (x_1 + x_3 - x_2 - x_4)^2 + (x_1 - x_3)^2 + (x_2 - x_4)^2$$

$$= 2 (x_1^2 + x_2^2 + x_3^2 + x_4^2 - x_1 x_2 - x_2 x_3 - x_3 x_4 - x_4 x_1)$$

$$= (x_1 - x_2)^2 + (x_2 - x_3)^2 + (x_3 - x_4)^2 + (x_4 - x_1)^2.$$

Similarly, we have

$$4 \left(\frac{y_1 + y_3}{2} - \frac{y_2 + y_4}{2} \right)^2 + (y_1 - y_3)^2 + (y_2 - y_4)^2$$

$$= (y_1 - y_2)^2 + (y_2 - y_3)^2 + (y_3 - y_4)^2 + (y_4 - y_1)^2.$$

Then, summing up and by the distance formula, we have

$$4|MN|^2 + |A_1 A_3|^2 + |A_2 A_4|^2 = |A_1 A_2|^2 + |A_2 A_3|^2 + |A_3 A_4|^2 + |A_4 A_1|^2.$$

Remark. This problem is solved by the analytic method. Generally speaking, the analytic method has the following rules.

If there are right angles in geometry problems, you can let two axes be on the two sides of the right angle, respectively, then use analytic geometry to solve the problems. For some problems, using (complex) numbers instead of letters does not affect the essence of the problem at all. Conversely, sometimes using letters instead of numbers reveals the intrinsic structure of the objects.

Of course, which choice of notation is more appropriate depends on the problem. In this problem, we choose the most general coordinates, so that it is easy to maintain the symmetry of the calculation, easy to observe and prove at the same time. Because of the symmetric relation between the horizontal and vertical coordinates, one proves the equation for the horizontal coordinate and at the same time obtains the equation for the vertical coordinate by symmetry.

Example 8. Two tangent lines of the circumcircle of an acute-angled triangle $\triangle ABC$ at A and B intersect at point D, and M is the midpoint of AB. Prove that $\angle ACM = \angle BCD$ (Problem from the China Training Team for 2017 International Mathematical Olympiad).

Solution. Represent points by complex numbers.

Let the circumcircle of $\triangle ABC$ be the unit circle on the complex plane with centre O and let the ray OM point to the direction of the real axis.

By the condition of the problem, point D is on the extended line of OM. $OM \cdot OD = OA^2 = 1$.

Let points A, B, C correspond to complex number $z, \bar{z}, c$, respectively. Then point M corresponds to Rez (see Fig. 10.1).

Since $OM \cdot OD = OA^2 = 1$, point D corresponds to $\frac{1}{\text{Re}z}$.

By the fact that $\angle ACM$ and $\angle BCD$ are acute, we need only to show that $H = \frac{c-z}{c-\frac{1}{\text{Re}z}} : \frac{c-\text{Re}z}{c-\bar{z}} \in \mathbb{R}$, that is, $\overline{H} = H$.

Denote $(c-z)(c-\bar{z}) = P$, $\left(c - \frac{1}{\text{Re}z}\right)(c - \text{Re}z) = Q$, then $H = \frac{P}{Q}$.

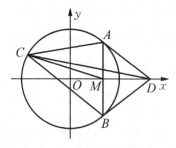

Figure 10.1

Note that $z \cdot \bar{z} = c \cdot \bar{c} = 1$, then

$$\bar{P} = (\bar{c} - \bar{z})(\bar{c} - z) = \left(\frac{1}{c} - \frac{1}{z}\right)\left(\frac{1}{c} - \frac{1}{\bar{z}}\right) = \frac{1}{c^2}(z - c)(\bar{z} - c) = \frac{P}{c^2},$$

$$\bar{Q} = \left(\frac{1}{c} - \frac{1}{\text{Re}z}\right)\left(\frac{1}{c} - \text{Re}z\right) = \frac{1}{c^2\text{Re}z}\left(\text{Re}z - c\right)\left(1 - c\text{Re}z\right)$$

$$= \frac{1}{c^2\text{Re}z} \cdot Q\text{Re}z = \frac{Q}{c^2},$$

Hence, $\bar{H} = \frac{\bar{P}}{\bar{Q}} = \frac{P}{Q} = H$. Therefore, $\angle ACM = \angle BCD$.

Example 9. There are n boxes $B_1, B_2, \ldots, B_n$ from left to right, and there are n balls in these boxes. Define an *operation* as follows: choose a box B_k, (1) If $k = 1$ and B_1 contains at least 1 ball, then move 1 ball from B_1 to B_2. (2) If $k = n$ and B_n contains at least 1 ball, then move 1 ball from B_n to B_{n-1}; (3) If $2 \leq k \leq n - 1$ and B_k contains at least 2 balls, then move 1 ball from B_k to B_{k+1} and 1 ball from B_k to B_{k-1}. Prove that for any arrangement of the n balls, we can achieve the state that each box has one ball after a finite times of *operations* (Problem from the 2011 China Girls' Mathematical Olympiad).

Solution. For any two vectors $\mathbf{x} = (x_1, x_2, \ldots, x_n)$ and $\mathbf{y} = (y_1, y_2, \ldots, y_n)$, if there is $1 \leq k \leq n$ such that $x_1 = y_1, \ldots, x_{k-1} = y_{k-1}, x_k > y_k$, then we denote $\mathbf{x} \succ \mathbf{y}$. We use a vector $\mathbf{x} = (x_1, x_2, \cdots, x_n)$ of non-negative integers to represent the placement of the balls in n boxes. Having an operation on B_k, the vector $\mathbf{x}$ changes to $\mathbf{x} + \alpha_k$, where $\alpha_1 = (-1, 1, 0, \ldots, 0)$, $\alpha_k = (\underbrace{0, \ldots, 0}_{k-2}, 1, -2, 1, 0, \ldots, 0)$, $(2 \leq k \leq n - 1)$, $\alpha_n = (0, \ldots, 0, 1, -1)$. Thus, if $k \geq 2$, $\mathbf{x} + \alpha_k \succ \mathbf{x}$. Hence, for any initial placement of the balls, we can have operations on $B_2, \ldots, B_n$ (as long as B_k contains at least two balls), such that the placement $\mathbf{y} = (y_1, y_2, \ldots, y_n)$ satisfies $y_k \leq 1, \forall k \geq 2$. If $y_2 = \cdots = y_n = 1$, the problem is proved. Otherwise, $y_1 \geq 2$. Let i be the least integer such that $y_i = 0$, having operations on $B_1, \ldots, B_{i-1}$ successively by

$$(y_1, 1, \ldots, 1, 0, y_{i+1}, \ldots, y_n)_{B_1, B_2, \ldots, B_{i-1}}$$

$$\succ (y_1, 1, \ldots, 1, 0, 1, y_{i+1}, \ldots, y_n)_{B_1, B_2, \ldots, B_{i-2}}$$

$$\succ (y_1, 1, \ldots, 1, 0, 1, 1, y_{i+1}, \ldots, y_n) \succ \cdots$$

$$\succ (y_1, 0, 1, \ldots, 1, y_{i+1}, \ldots, y_n)_{B_1} \succ (y_1 - 1, 1, \ldots, 1, y_{i+1}, \ldots, y_n).$$

Repeatedly performing the above operations, the desired state will be achieved at last.

Exercises

1. Mr. Lee had a six-digit home phone number. It was first upgraded to a seven-digit number by inserting 8 between the first and the second digit. Then it was upgraded again to an eight-digit number by putting 2 ahead of the first digit. Mr. Lee found out that his eight-digit number is 81 times his original six-digit phone number. What was Mr Lee's original phone number?

2. Find the number of non-empty subsets of the set $\{1, 2, 3, \ldots, 2009\}$ whose sum of elements is odd.

3. Prove that there exist infinitely many pairs (m, n) of positive integers satisfying

$$m^2 + 25n^2 = 10mn + 7(m + n).$$

4. Determine all positive integers n with the property that $n = (d(n))^2$. Here $d(n)$ denotes the number of positive divisors of n) (Problem from the 1999 Canadian Mathematical Olympiad).

5. For given real number x, find the value of the expression $S = \sum_{n=1}^{\infty} \frac{(-1)^{\lfloor 2^n x \rfloor}}{2^n}$.

6. If $a, b, c, x, y, z \in \mathbb{R}$, satisfy

$$\begin{cases} \cos(x + y + z) = a \cos x + b \cos y + c \cos z, \\ \sin(x + y + z) = a \sin x + b \sin y + c \sin z, \end{cases}$$

   prove that $\begin{cases} a \cos(y + z) + b \cos(z + x) + c \cos(x + y) = 1, \\ a \sin(y + z) + b \sin(z + x) + c \sin(x + y) = 0. \end{cases}$

7. Make isosceles right-angled triangles $\triangle ABE$, $\triangle BCF$, $\triangle CDG$, $\triangle DAH$ outside the quadrilateral $ABCD$, with AB, BC, CD, DA as the hypotenuses, respectively. Prove that $EG \perp FH$ and $EG = FH$.

8. Let $\triangle ABC$ be anacute-angled triangle with orthocentre H. The tangent lines from A to the circle with diameter BC touch this circle at P and Q. Prove that P, H, Q are collinear (Problem from the 1996 China Mathematical Olympiad).

9. For each positive integer n, $f_n = \lfloor 2^n \cdot \sqrt{2008} \rfloor + \lfloor 2^n \cdot \sqrt{2009} \rfloor$, prove that there are infinitely many odd numbers and infinitely many even numbers in the sequence $\{f_n\}$ (Problem from the 2008 China Girls' Mathematical Olympiad).

Chapter 11

Combine Numbers and Figures

The combination of numbers and figures is a very important idea and a significant strategy in problem solving.

Numbers and figures reflect the attributes of two aspects of things. The combination of numbers and figures is to solve mathematical problems through correspondence and transformation between them. To be specific, in the process of problem solving, the geometric properties are concretely quantified by the deduction of quantitative relations, and the quantitative relations are visualized by the geometric background. This strategy can simplify the complex and reify the abstract, and help one to grasp the essence of mathematical problems quickly and solve them easily.

Example 1. Suppose that a quadratic equation $x^2 + ax + 2b - 2 = 0$ of real coefficients has two real roots, one in $(0, 1)$, and the other in $(1, 2)$. Find the range of $\frac{b-4}{a-1}$ (Problem from the 2008 Zhejiang Province High School Mathematical Contest).

Solution. Since $f(x) = x^2 + ax + 2b - 2$ has one root in $(0, 1)$ and one root in $(1, 2) \Leftrightarrow f(0) > 0$, $f(1) < 0$, $f(2) > 0$. That is,

$$\begin{cases} b > 1, \\ a + 2b < 1, \\ a + b > -1. \end{cases}$$

Draw the region (in black) at plane aOb in which the above inequalities hold, as shown in Fig. 11.1. Observe that $\frac{b-4}{a-1}$ is the slope of the line from point $P(1, 4)$ to point (a, b) in the region, and by Fig. 11.1, we see that the maximal slope $3/2$ is that of the line passing through points $P(1,4)$

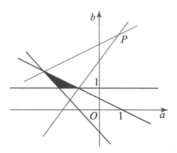

Figure 11.1

and $(-1,1)$, and the minimal slope $1/2$ is that of the line passing through points P $(1, 4)$ and $(-3, 2)$. So, $\frac{b-4}{a-1} \in \left(\frac{1}{2}, \frac{3}{2}\right)$.

Remark. Let us see another solution for this problem.

Let the two distinct roots be x_1, x_2, and $0 < x_1 < 1 < x_2 < 2$. Then

$$1 < x_1 + x_2 = -a < 3, \quad 0 < x_1 x_2 = 2b - 2 < 2.$$

That is, $-3 < a < -1$, $1 < b < 2$. So $\frac{1}{4} < \frac{1}{1-a} < \frac{1}{2}$, $2 < 4 - b < 3$.

Multiply both sides of the above two inequalities, to yield, $\frac{1}{2} < \frac{b-4}{a-1} < \frac{3}{2}$.

Is this solution correct? If not, why does it give the correct answer?

Example 2. Two circles O_1 and O_2 intersect at two points A and B. CD is passing through A and perpendicular to AB with C on circle O_1 and D on circle O_2. MC and ND are perpendicular to CD, such that $MC = MB$ and $ND = NB$. PQ is passing through B and parallel to MN with P on circle O_1 and Q on circle O_2. Prove that $BP = BQ$.

Solution. The following is a proof by analytic geometry.

Without loss of generality, suppose that the two circles $O_1(x_1, 0)$ and $O_2(x_2, 0)$, with $x_1 < 0 < x_2$, intersect at points $A(0, -1)$ and $B(0, 1)$. Since $CD \perp AB$, we determine the positions $C(2x_1, -1)$ and $D(2x_2, -1)$. Denote $M = M(2x_1, y_1)$ and $N = N(2x_2, y_2)$, then $MC^2 = BM^2 \Rightarrow (y_1 + 1)^2 = 4x_1^2 + (y_1 - 1)^2 \Rightarrow y_1 = x_1^2$. Similarly, $y_2 = x_2^2$. Thus, the slope of MN is $\frac{y_2 - y_1}{2(x_2 - x_1)} = \frac{x_1 + x_2}{2}$. Denote $P = P(x, y)$, $Q = Q(x, y)$. Since $PQ \parallel MN$ and PQ passes through B, $y = \frac{x_1 + x_2}{2}x + 1$. Further, since P is on circle $O_1(x_1, 0)$: $(x - x_1)^2 + y^2 = x_1^2 + 1$, we obtain $x_p = \frac{x_1 - x_2}{1 + (x_1 + x_2)^2/4}$. Interchanging x_1 and x_2, we obtain $x_q = -x_p$. That is, the projected lengths of BP and BQ on x-axis are equal, so, $BP = BQ$.

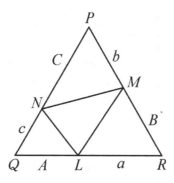

Figure 11.2

Example 3. Suppose that positive real numbers a, b, c, A, B, C satisfy $a + A = b + B = c + C = k$. Prove that

$$aB + bC + cA < k^2.$$

Solution. Construct an equilateral triangle $\triangle PQR$ with side length k (see Fig. 11.2). Take L, M, N on the sides such that $QL = A$, $RM = B$, $PN = C$, so $LR = a$, $MP = b$, $NQ = c$. Then

$$S_{\triangle LRM} = \tfrac{1}{2}aB \sin 60°, \quad S_{\triangle MPN} = \tfrac{1}{2}bC \sin 60°,$$

$$S_{\triangle NQL} = \tfrac{1}{2}cA \sin 60°, \quad S_{\triangle PQR} = \tfrac{1}{2}k^2 \sin 60°.$$

Obviously, by Fig. 11.2, we have

$$S_{\triangle LRM} + S_{\triangle MPN} + S_{\triangle NQL} < S_{\triangle PQR}.$$

Hence,

$$\frac{1}{2}aB \sin 60° + \frac{1}{2}bC \sin 60° + \frac{1}{2}cA \sin 60° < \frac{1}{2}k^2 \sin 60°.$$

That is

$$aB + bC + cA < k^2.$$

Example 4. Let p and q be coprime. Prove that

$$\left\lfloor \frac{p}{q} \right\rfloor + \left\lfloor \frac{2p}{q} \right\rfloor + \cdots + \left\lfloor \frac{(q-1)p}{q} \right\rfloor = \frac{(p-1)(q-1)}{2}.$$

Solution. Consider a rectangle $OABC$ on a Cartesian plane with vertices $O(0,0)$, $A(q, 0)$, $B(q, p)$, $C(0, p)$. Since p and q are coprime, $y = \frac{p}{q}x$ is not an integer for integer $x \in (0, q)$. That is, there are no lattice points on segment OB other than O and B.

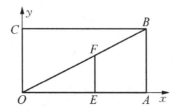

Figure 11.3

In the following, we count the number s of lattice points in the interior of $\triangle OAB$ in two ways.

On the one hand, make a line EF perpendicular to x-axis passing througha lattice point $E(k,0)\,(0 < k < q)$ on x-axis and intersect OB at F (see Fig. 11.3). The ordinate of F is $y = \frac{kp}{q}$. Thus, there are $\left\lfloor \frac{kp}{q} \right\rfloor$ lattice points in the interior of segment EF. Hence,

$$s = \left\lfloor \frac{p}{q} \right\rfloor + \left\lfloor \frac{2p}{q} \right\rfloor + \cdots + \left\lfloor \frac{(q-1)p}{q} \right\rfloor.$$

On the other hand, there are $(p-1)(q-1)$ lattice points in the interior of rectangle $OABC$, and there are no lattice points in the interior of segment OB. Due to symmetry, the number of lattice points in the interior of $\triangle OAB$ and of $\triangle BCO$ are the same. Therefore,

$$s = \frac{(p-1)(q-1)}{2}.$$

The combination of the above equations completes the proof.

Remark. A short proof of Example 4 is as follows:

Let r be a positive integer $(1 \le r \le q-1)$, then we have

$$\left\lfloor \frac{(q-r)p}{q} \right\rfloor = p - \left\lfloor \frac{rp}{q} \right\rfloor - 1.$$

Thus, we have the following conclusion:

$$\left\lfloor \frac{(q-1)p}{q} \right\rfloor + \cdots + \left\lfloor \frac{2p}{q} \right\rfloor + \left\lfloor \frac{p}{q} \right\rfloor = (p-1)(q-1)$$

$$- \left(\left\lfloor \frac{p}{q} \right\rfloor + \left\lfloor \frac{2p}{q} \right\rfloor + \cdots + \left\lfloor \frac{(q-1)p}{q} \right\rfloor \right).$$

Though it is simpler, the former proof is intuitive and suitable for many problems involving lattice points.

Example 5. On Sundays, a cinema has a children's show for only 5 Yuan per child. Each child is limited to buying one ticket. The booking office has no money for change at the beginning when $2n$ children stand in a queue to buy tickets. Half of them have only a 5 Yuan note each, and the others have only a 10 Yuan note. The booking clerk can start selling tickets to anyone in the queue to the end in the order, then from the beginning to the rest in the order. The question is, can the clerk sell all $2n$ tickets without any trouble in terms of shortage of change at any moment?

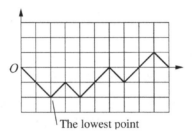

Figure 11.4

Analysis Take a grid paper with x and y axes and investigate the notes that children have from the beginning of the queue. Begin at the origin, write a '/' in the grid if one gets a 5 Yuan note or a '\' if one gets a 10 Yuan note. Thus, a polyline is drawn with endpoints on the x-axis.

Find the lowest point on the polyline. If there are several, take one that is the nearest to the y-axis, say, the point has abscissa k. Then selling tickets from the $(k+1)$th kid will do. Let us show it by an example.

Let the 10 kids in the queue have their notes in order as 10, 10, 5, 10, 5, 5, 10, 5, 5, 10. Then the polyline is as shown in Fig. 11.4.

The reader may check that the clerk can start from the third kid.

Remark. Another similar problem:

There are n petrol stations in a circular road. The total amount of petrol is just enough for a car to finish one circuit. Prove that there is a station at which a car with an empty tank can refuel and start and complete the entire journey.

Example 6. Prove that for any eight real numbers $a_1, a_2, \ldots, a_8$, among the following six numbers $a_1a_3 + a_2a_4, a_1a_5 + a_2a_6, a_1a_7 + a_2a_8, a_3a_5 + a_4a_6,$ $a_3a_7 + a_4a_8, a_5a_7 + a_6a_8$, at least one of them is non-negative. And give an example in which just one is non-negative.

Solution. Consider four points $A(a_1, a_2), B(a_3, a_4), C(a_5, a_6), D(a_7, a_8)$ on a Cartesian plane with O as the origin.

If there is a point, say A, that coincides with O, then $a_1 a_3 + a_2 a_4 = 0$ is non-negative.

If no points in A, B, C, D coincide with O, without loss of generality, suppose that rays OA, OB, OC, OD are in anticlockwise order. By the drawer principle, at least one angle of $\angle AOB, \angle BOC, \angle COD, \angle DOA$ is $\leq 90°$. Say, $0 \leq \angle AOB \leq 90°$, then $a_1 a_3 + a_2 a_4 = \overrightarrow{OA} \cdot \overrightarrow{OB} \geq 0$.

An example is $(a_1, a_2, a_3, a_4, a_5, a_6, a_7, a_8) = (3, -1, 3, 1, -1, 2, -1, -2)$, where $0 < \angle AOB < 90°$, and other angles $> 90°$, so, only $a_1 a_7 + a_2 a_8 = 8 > 0$.

Example 7. How many rays at most, all emanating from a point in space, satisfy that the angle formed by any two rays is obtuse?

Solution. First, take rays from the centre of a regular tetrahedron to its four vertices. The angles formed by any two of these rays are obtuse.

Next, we show that five rays cannot satisfy the condition of the problem.

Suppose, on the contrary there are five rays $OA_1, OA_2, \ldots, OA_5$ emanating from point O such that the angle formed by any two of them is obtuse.

Let $\overrightarrow{OA_5} = (0, 0, -1)$, and $\overrightarrow{OA_i} = (x_i, y_i, z_i)(i = 1, 2, 3, 4)$.

By $\overrightarrow{OA_i} \cdot \overrightarrow{OA_5} = -z_i$, we see that $z_i > 0 \ (i = 1, 2, 3, 4)$.

Denote by P_i the projection of A_i onto the xOy plane.

By the drawer principle, in four vectors $\overrightarrow{OP_i} = (x_i, y_i, 0) \ (i = 1, 2, 3, 4)$, there is an angle $\leq 90°$ formed by two vectors, say $\overrightarrow{OP_1}$ and $\overrightarrow{OP_2}$. So,

$$\overrightarrow{OP_1} \cdot \overrightarrow{OP_2} = x_1 x_2 + y_1 y_2 \geq 0.$$

Thus,

$$\overrightarrow{OA_1} \cdot \overrightarrow{OA_2} = x_1 x_2 + y_1 y_2 + z_1 z_2 > 0.$$

That is, the angle formed by $\overrightarrow{OA_1}$ and $\overrightarrow{OA_2}$ is acute. A contradiction.

Example 8. Let a, b, c, d be integers with $a > b > c > d > 0$. Suppose that

$$ac + bd = (b + d + a - c)(b + d - a + c).$$

Prove that $ab + cd$ is not prime (Problem from the 2001 International Mathematical Olympiad).

Solution. By $ac + bd = (b + d + a - c)(b + d - a + c) = (b + d)^2 - (a - c)^2$, we see that $a^2 - ac + c^2 - 2ac \cos 60° = b^2 + d^2 - 2bd \cos 120°$. We can construct a quadrilateral $ABCD$ such that $DA = a$, $AB = a$, $\angle A = 60°$, and $BC = d, CD = b$, $\angle C = 120°$. So, A, B, C, D are consyclic. Then let $\angle D = \alpha$, by the law of cosine, we have,

$$BD^2 = a^2 - ac + c^2,$$

$$AC^2 = a^2 + d^2 - 2ad \cos \alpha = b^2 + c^2 - 2bc \cos(180° - \alpha)$$

$$= b^2 + c^2 + 2bc \cos \alpha. \tag{1}$$

Solving it yields $2 \cos \alpha = \frac{a^2 + d^2 - b^2 - c^2}{ad + bc}$, hence,

$$AC^2 = a^2 + d^2 - ad \cdot \frac{a^2 + d^2 - b^2 - c^2}{ad + bc}$$

$$= \frac{(a^2 + d^2)(ad + bc) - ad(a^2 + d^2 - b^2 - c^2)}{ad + bc}$$

$$= \frac{(a^2 + d^2)bc + ad(b^2 + c^2)}{ad + bc} = \frac{(ab + cd)(ac + bd)}{ad + bc}. \tag{2}$$

By Ptolemy's theorem, $AC \cdot BD = a \cdot b + c \cdot d$, taking squares of both sides, we have

$$AC^2 \cdot BD^2 = (ab + cd)^2. \tag{3}$$

Substitute (1) and (2) into (3), and rearrange the terms to yield,

$$(a^2 - ac + c^2)(ac + bd) = (ab + cd)(ad + bc).$$

Hence,

$$(ac + bd) | (ab + cd)(ad + bc). \tag{4}$$

Since $a > b > c > d$, we have $ab + cd > ac + bd > ad + bc$. If $ab + cd$ is prime, then $ac + bd$ and $ab + cd$ are coprime. By (4), we see that $(ac + bd) | ad + bc$, contradicting $ac + bd > ad + bc$! Therefore, $ab + cd$ is not prime.

Remark. In fact, there is a pure algebraic simple proof of (4) as follows.
Since $a^2 - ac + c^2 = b^2 + bd + d^2 = M$, then,

$$\frac{(ab + cd)(ad + bc)}{ac + bd} = \frac{(a^2 + c^2)bd + (b^2 + d^2)ac}{ac + bd}$$

$$= \frac{(a^2 - ac + c^2)bd + (b^2 + bd + d^2)ac}{ac + bd}$$

$$= \frac{Mbd + Mac}{ac + bd} = M.$$

This gives (4) and the rest is the same. However, the original prove is more natural.

It can be seen that 'number' and 'figure' in the solution of Example 8 are transforming and influencing each other, and they finally optimize the process of problem solving. Just as the famous Chinese mathematician Hua Luogeng once wrote a poem for combining of 'number' and 'figure':

Numbers and figures depend on each other, how can they get away from each other.

When numbers lack a figure, there is little intuition and when shape lacks a number, there is little precision.

Number and figure thus stand when united and fall when divided, divided they fall.

Geometry and algebra should always connect and never separate.

Exercises

1. Prove that for any real number x,
$$\left| \sqrt{x^2 + x + 1} - \sqrt{x^2 - x + 1} \right| < 1.$$

2. The film ticket of a cinema is 5 Yuan per person. Suppose that there is no money for change in the booking office at the beginning. There are 10 people in a queue for booking tickets, 5 of them have 5 Yuan notes each, others have 10 Yuan notes each. If people queue up in a random order, find the probability of selling tickets without trouble in giving change (Problem from the 2010 Hebei Province High School Mathematical Contest).

3. In the quadratic equation $x^2 + z_1 x + z_2 + m = 0$ in x, z_1, z_2, m are all complex numbers, and $z_1^2 - 4z_2 = 16 + 20i$. Let the two roots of the quadratic equation α, β satisfy $|\alpha - \beta| = 2\sqrt{7}$. Find the maximum and the minimum of $|m|$ (Problem from the 1994 China High School Mathematical League).

4. Suppose that the graph of the function $f(x) = |\sin x|$ and the line $y = kx$ ($k > 0$) have exactly three intersection points, the maximal abscissa of which is α. Prove that $\frac{\cos \alpha}{\sin \alpha + \sin 3\alpha} = \frac{1 + \alpha^2}{4\alpha}$ (Problem from the 2008 China High School Mathematical League).

5. Positive numbers x, y, and z satisfy the following system:
$$\begin{cases} x^2 + xy + y^2/3 = 25, \\ y^2/3 + z^2 = 9, \\ x^2 + xz + z^2 = 16. \end{cases}$$

Find the value of the expression $xy + 2yz + 3zx$ (Problem from the 1984 All-Soviet Union Mathematical Olympiad).

6. Let $\alpha \geq 0$ be given and z_1, z_2 be complex numbers. Find the smallest real number $\lambda = \lambda(\alpha)$, such that $|z_1 - xz_2| \leq \lambda|z_1 - z_2|$ for all $x \in [0, 1]$ subject to condition $|z_1| \leq \alpha|z_1 - z_2|$ (Problem from the 2011 China Girls Mathematical Olympiad).

7. In an orthogonal coordinate system in space, any three points of the origin O and points $A, B_1, B_2, \ldots, B_6$ are not collinear. Point $A(a, b, c)$ satisfies $a + b + c = 0$, and the coordinates of points $B_1, B_2, \ldots, B_6$ are (x_1, x_2, x_3), (x_1, x_3, x_2), (x_2, x_1, x_3), (x_2, x_3, x_1), (x_3, x_1, x_2), (x_3, x_2, x_1), respectively. Prove that there are at least two acute angles in $\angle AOB_i$ $(i = 1, 2, \ldots, 6)$.

8. A point P is within a triangle $\triangle ABC$ with $BC = a$, $CA = b$, $AB = c$. Prove that

(a) $a \cdot PB \cdot PC + b \cdot PC \cdot PA + c \cdot PA \cdot PB \geq abc$;

(b) $a \cdot PA^2 + b \cdot PB^2 + c \cdot PC^2 \geq abc$.

Chapter 12

Correspondence and Pairing

Correspondence is a basic and important mathematical concept in problem solving. The idea is roughly like this: if a problem can be transformed into another problem by a one-to-one correspondence, such that the latter problem can be solved easily, then the former problem is solved by the inverse correspondence. Professor Xu Lizhi called this method Relationship, Mapping, and Inversion (RMI). Pairing is a method of grouping objects in pairs by some correspondence so as to simplify the problem.

Example 1. The rule of the game *24 points* is to randomly give out four positive integers not greater than 10, then use four arithmetic operations on the four numbers such that the result is 24. How many cases are there in the game *24 points*?

Solution 1. (I) If four numbers are the same, as (x, x, x, x), then there are 10 cases.

(II) If there are two different numbers, as (x, x, x, y), $x \neq y$, then there are $10 \times 9 = 90$ cases; as (x, x, y, y), $x \neq y$, there are $\binom{10}{2} = 45$ cases. So, there are 135 cases in total.

(III) If there are three different numbers, as (x, x, y, z). There are 10 ways to make the value of x. Then select another for y and z from the remaining nine numbers, there are $\binom{9}{2} = 36$ ways, by the multiplication principle, so there are 360 cases in total.

(IV) If four numbers are distinct, as (x, y, z, w), then there are $\binom{10}{4} = 210$ cases.

Summing up, there are $10 + 135 + 360 + 210 = 715$ cases in total.

Solution 2. Let the four numbers be an ordered four-element array (a, b, c, d), $(1 \leq a \leq b \leq c \leq d \leq 10)$. Then the problem is converted to finding the number of such arrays.

Define a mapping $f : (a, b, c, d) \mapsto (a, b+1, c+2, d+3)$, we see that the mapping f is a bijection of

$$X = \{(a, b, c, d) | 1 \leq a \leq b \leq c \leq d \leq 10\}$$

onto the set

$$Y = \{(a', \ b', \ c', \ d') | 1 \leq a' < b' < c' < d' \leq 13\}.$$

So, $|X| = |Y|$.

Obviously, the number of elements of Y is the number of four-element arrays in the set $\{1, 2, \ldots, 13\}$, which is $\binom{13}{4} = 715$.

Remark. Solution 2 is more insightful than Solution 1 by use of the correspondence method.

Example 2. The rule of the challenge series of Go is that there are two teams A and B, with the same number of players and each team is pre-ordered. First, two No. 1 players play, the loser leaves the game and the winner plays with No. 2 of the opponent team, $\ldots$, when the last player of a team loses, then the opponent team wins.

Find the number of all possible processes of the tournament if each team has seven members (Problem from the 1988 China High School Mathematical League).

Solution 1. In each tournament, let the losers queue up in the order of time, followed by the winner and those who did not play as per the pre-order. It is easy to see that each sequence corresponds to a tournament process. So, there are in all $\binom{14}{7} = 3432$ ways of queuing. And 3432 possible processes of the tournament in total.

Solution 2. Let the ith player of Team A win over $x_i (1 \leq i \leq 7)$ person(s) of B. It is easy to see that, each tournament process such that Team A wins corresponds to a non-negative integer solution $(x_1, x_2, \ldots, x_7)$ of $x_1 + x_2 + \cdots + x_7 = 7$. Since the number of non-negative integer solutions is $\binom{7+6}{6} = \binom{13}{6} = 1716$, so is the number of the cases in which Team A wins. Similarly, the number of the cases in which Team B wins is also 1716. So, there are 3432 in total.

Remark. Both solutions use the correspondence method. Solution 1 is easier. Solution 2 uses the result that the number of non-negative integer solutions of $x_1 + x_2 + \cdots + x_n = m$ is $\binom{m+n-1}{n-1}$, which can also be proved by the correspondence method.

Example 3. The sum of the maximum and the minimum of a number set A is called the *characteristic* of the set A, and denoted by $m(A)$. Find the average of *characteristics* of all non-empty subsets of $X = \{1, 2, \ldots, n\}$.

Solution. Let Y be the set of all non-empty subsets of X. For $A \in Y$, let
$$A' = \{n + 1 - a | a \in A\},$$
then $A' \in Y$. Thus $f : A \mapsto A'$ is a bijection of set Y onto Y itself.

The average of the *characteristics* is
$$g = \frac{1}{|Y|} \sum_{A \in Y} m(A) = \frac{1}{2|Y|} \sum_{A \in Y} m(A) + m(A').$$
Since the maximum of A is the minimum of A', and their sum is $n + 1$; the minimum of A is the maximum of A', their sum is also $n + 1$. Thus,
$$g = \frac{1}{2|Y|} \sum_{A \in Y} 2(n+1) = (n+1) \cdot \frac{1}{|Y|} \sum_{A \in Y} 1 = n + 1.$$

Remark. In the solution, the pairing (A, A') is used.

Example 4. Let a, b, c be integers each with absolute value less than or equal to 10. The cubic polynomial $f(x) = x^3 + ax^2 + bx + c$ satisfies the property $|f(2 + \sqrt{3})| < 0.0001$. Determine whether $2 + \sqrt{3}$ is a root of f (Problem from the 2007 China Girls' Mathematical Olympiad).

Solution. Substitute $2 + \sqrt{3}$ into the polynomial,
$$f(2 + \sqrt{3}) = (2 + \sqrt{3})^3 + a(2 + \sqrt{3})^2 + b(2 + \sqrt{3}) + c$$
$$= (26 + 7a + 2b + c) + (15 + 4a + b)\sqrt{3}.$$
Denote $26 + 7a + 2b + c = m$, $15 + 4a + b = n$, then
$$|m + n\sqrt{3}| = |f(2 + \sqrt{3})| < 0.0001,$$
where $m, n \in \mathbb{Z}$, and $|m| \leq 126, |n| \leq 65$. Thus,
$$|m - n\sqrt{3}| \leq |m| + |n|\sqrt{3} < |m| + 2|n| \leq 256.$$
Hence,
$$|m^2 - 3n^2| = |m + n\sqrt{3}| \cdot |m - n\sqrt{3}| < 0.0001 \times 256 < 1.$$
Since $m^2 - 3n^2$ is an integer, $m^2 - 3n^2 = 0$. It must be $m = n = 0$, $f(2 + \sqrt{3}) = 0$, that is, $2 + \sqrt{3}$ is a root of the polynomial.

Remark. In the solution, the conjugate pair $(m+n\sqrt{3},\ m-n\sqrt{3})$ is used. If $\left|m^2 - 3n^2\right| < 1$, then $m^2 - 3n^2 = 0$.

Example 5. For $n \in \mathbb{N}^+$, denote the product of all positive integers $\leq n$ and coprime to n by $\pi(n)$. Prove that $\pi(n) \equiv \pm 1 \pmod{n}$.

Solution. If $n = 1, 2$, the conclusion is surely true. Let $n \geq 3$ and X_n be the set of all positive integers $\leq n$ that are coprime to n. First, we prove a lemma.

Lemma. *For any $t \in X_n$, there is a unique $t' \in X_n$, such that $t' \cdot t \equiv 1 \pmod{n}$.*

Proof. Since t, n are coprime, there exist integers p, q such that $pt + qn = 1$ with p, n coprime. Take $t' \in X_n$ such that $t' \equiv p \pmod{n}$. We see that $t' \cdot t \equiv 1 \pmod{n}$. If $t'_1 \cdot t \equiv t'_2 \cdot t \equiv 1 \pmod{n}$, then $t'_1 \equiv t'_1 \cdot t \cdot t'_2 \equiv t'_2 \pmod{n}$. Thus, such $t' \in X_n$ is unique. The lemma is proved.

(I) If $t \neq t'$, $t' \cdot t \equiv 1 \pmod{n}$, put t and t' in a group. Suppose that there are i such groups in total.

(II) If $t = t'$, then $t^2 \equiv 1 \pmod{n}$. So, $(n - t)^2 \equiv n^2 - 2tn + t^2 \equiv 1 \pmod{n}$, that is, $n - t = (n - t)'$. Put t and $n - t$ in a group (where $t \neq n - t$, since if n is odd, $t \neq n - t$, and if n is even, $t \neq \frac{n}{2}$). Suppose that there are j such groups, each with $t(n - t) \equiv -t^2 \equiv -1 \pmod{n}$.

By (I) and (II), taking products of all pairs in X_n, we have

$$\pi(n) \equiv 1^i \cdot (-1)^j \equiv \pm 1 \pmod{n}.$$

$\square$

Remark. The solution is derived using pairing of inverse and opposite numbers in modulo n. In the following, we give a more in-depth discussion on $\pi(n)$.

For $t \in X_n$, define the order of t about n by $r_n(t) = r$, where r is the least positive integer such that $t^r \equiv 1 \pmod{n}$. Thus, if $t = t' \neq 1$, then $r_n(t) = 2$. And $r_n(1) = 1$. The solution of Problem 5 reveals the fact: for $n \geq 3$, if the number m of elements of order 2 in X_n satisfies $m \equiv 1 \pmod{4}$, then $\pi(n) \equiv -1 \pmod{n}$; if $m \equiv 3 \pmod{4}$, then $\pi(n) \equiv 1 \pmod{n}$. The following are two corollaries:

Corollary 1 (Wilson's theorem). *If p is prime, then $(p - 1)! \equiv -1 \pmod{p}$. In fact, we can check directly for $p = 2$. For any odd prime p, $p - 1$ is the unique element of order 2 in X_p, thus $\pi(p) = (p - 1)! \equiv -1 \pmod{p}$.*

Corollary 2. *If n is even, then $\pi(2n) \equiv 1 \pmod{n}$.*

In fact, in this case, $(i, 2n) = 1$ and $(i, n) = 1$ are equivalent, and $n + i \equiv i \pmod{n}$, thus $\pi(2n) \equiv \pi^2(n) \equiv 1 \pmod{n}$.

A result stronger than Corollary 2 is if $4|n$, $n \geq 8$, then $\pi(n) \equiv 1 \pmod{n}$.

In fact, for each element $t \in X_n$ of order 2, $1 \leq t < \frac{n}{4}$. Put four distinct numbers $t, \frac{n}{2} - t, \frac{n}{2} + t, n - t$ in a group, and their product $\equiv 1 \pmod{n}$.

The following two examples further demonstrate the technique of the correspondence method.

Example 6. There are $2n+1$ trees, $T_1, T_2, \ldots, T_{2n+1}$, on a straight road. A drunk starts his random walk from the tree T_{n+1} in the middle for n minutes following the rule: if he arrives at tree T_i ($2 \leq i \leq 2n$) at the end of the mth minute, then he will arrive at trees T_{i-1}, T_i, T_{i+1} with probabilities $\frac{1}{4}, \frac{1}{2}, \frac{1}{4}$, respectively, at the end of $(m + 1)$th minute. Find the probability p_i that the drunk arrives at tree T_i ($1 \leq i \leq 2n + 1$) at the end of the nth minute.

Solution. Suppose that $2n + 1$ trees $T_1, T_2, \ldots, T_{2n+1}$ line up from left to right with unit spacing. Define *a random walk* by walking $1/2$ unit to the left or the right with probability $1/2$. Thus, the drunk's move per minute is the result of two random walks. So, the problem is equivalent to finding the probability p_i of the drunk from T_{n+1} to T_i after $2n$ random walks.

For fixed i ($1 \leq i \leq 2n + 1$), suppose that the drunk arrives at T_i after k right walks and $2n - k$ left walks. Then

$$n + 1 + \frac{k - (2n - k)}{2} = i.$$

That is, $k = i - 1$ right walks, and $2n - (i - 1)$ left walks. Thus

$$p_i = \frac{1}{2^{2n}} \binom{2n}{i - 1}.$$

Example 7. If a sequence $\{a_n\}$ satisfies $\sum_{d|n} a_d = 2^n$, $\forall n \in \mathbb{N}^+$, prove that $n | a_n$.

Solution. It is easy to see that a_n is uniquely determined by the condition. Call a 0-1 sequence with length n *repeating* if the sequence has n terms, and is made by repeating its first d terms n/d times, $d < n$ and $d|n$. Then the sequence is made by repeating its first d terms Call the maximal possible value of $c = \max_{d|n} \left(\frac{n}{d}\right)$ the repeating times. Call other 0–1 sequences *non-repeating*, and the repeating time is 1. Thus, each 0–1 sequence with length n has the repeating times c, and $c|n$.

Let S be the set of 0–1 sequences with length n. The number of elements in S is 2^n. Classify every sequence in S by its repeating times. Then each sequence in S with repeating times c has its first $d = n/c$ terms. Let the number of sequences of length d be b_d, then b_d equals the number of 0–1 sequences with length n and repeat times $c = n/d$. Summing up for all $d \mid n$, we have $\sum_{d \mid n} b_d = 2^n$, especially, $b_1 = 2 = a_1$, and since $\{a_n\}$ and $\{b_n\}$ have the same recurrence relation, we have $b_n = a_n, \forall n \in \mathbb{N}^+$.

Note that, by circularly shifting, each unrepeated sequence with length n corresponds to n distinct unrepeated sequences, thus $n \mid b_n$, consequently $n \mid a_n$.

Example 8. Let a_n, $n \in \mathbb{N}^+$, be the number of such positive integers N: the digits of N can only be 1, or 3, or 4, and the sum of digits of N is n. Prove that a_{2n} is a perfect square (Problem from the 1991 China High School Mathematical League).

Solution. Denote A the set of positive integers N whose digits are 1, 3, or 4. $A_n = \{N \in A \mid$ the sum of digits of N is $n\}$, then $|A_n| = a_n$.

Denote B the set of positive integers N whose digits consist of 1, 2. $B_n = \{N \in B \mid$ the sum of digits of N is $n\}$. Let $|B_n| = b_n$. We shall show that $a_{2n} = b_n^2$.

Define a mapping $f : B \to A$, for $N \in B$, $f(N)$ is the number obtained as follows: from the left to the right of the digits of N, substitute 2 and the digit it follows by the sum of these two numbers successively (e.g. $f(1221212) = 14132$, $f(21121221) = 31341$). Thus,

$$f(B_{2n}) = A_{2n} \cup A'_{2n-2},$$

where $A'_{2n-2} = \{10k + 2 \mid k \in A_{2n-2}\}$. Hence,

$$b_{2n} = a_{2n} + a_{2n-2}.$$

Since $b_{2n} = b_n^2 + b_{n-1}^2$ (this is because a number in B_{2n} is made by either joining two numbers in B_n, or by inserting 2 between two numbers in B_{n-1}), hence,

$$a_{2n} + a_{2n-2} = b_n^2 + b_{n-1}^2, n \geq 2.$$

Since $a_2 = b_1^2 = 1$, we see that, for any $n \in \mathbb{N}$, we have $a_{2n} = b_n^2$.

Exercises

1. Select n points on a circle and link any two points on them by a chord. If any three chords have no common intersection points except on the circle, how many intersecting points exist in the interior of the circle?

2. For each nonempty subset of set $\{1, 2, \ldots, n\}$ ($n \in \mathbb{N}^+$), we define its *alternative sum*: put the elements of the set in the decreasing order, and alternatively add and subtract the numbers starting from the left to the right (e.g., the *alternative sum* of the set $\{9, 6, 4, 2, 1\}$ is $9 - 6 + 4 - 2 + 1 = 6$, and the *alternative sum* of the set $\{5\}$ is 5). Find the sum of all *alternative sums*.

3. Let positive integers r, n satisfy $1 \leq r \leq n$. Find the arithmetic mean $f(r, n)$ of all the minimums of r-element subsets of $\{1, 2, \ldots, n\}$.

4. Let A be a subset of the set $X = \{1, 2, \ldots, n\}$. If the sum of all elements in A is odd, then call A an *odd subset* of X. If it is even, then call it an *even subset*.

   (1) Find the numbers of *odd and even subsets* of X.
   (2) Find the sum of elements of all *odd subsets* for $n \geq 3$.

5. Let $n \equiv 1 \pmod 4$ and $n > 1$, $P = (a_1, a_2, \ldots, a_n)$ be any permutation of $(1, 2, \ldots, n)$, k_P be the largest subscript k such that $a_1 + a_2 + \cdots + a_k < a_{k+1} + a_{k+2} + \cdots + a_n$. Find the sum of k_P for all permutations.

6. Let $n > 2$ be a positive integer. Prove that the sum of cubics of the numbers in $\{1, 2, \ldots, n\}$ that are coprime to n is divisible by n.

7. For prime $p \geq 3$, prove that $\sum_{k=1}^{p-1} k^{k^2 - k + 1}$ is divisible by p.

8. Let k be a given positive integer. Prove that if $A = k + \frac{1}{2} + \sqrt{k^2 + \frac{1}{4}}$, then for any positive integer n, $\lfloor A^n \rfloor$ is divisible by k.

9. Let S be a finite set of points in three-dimensional space. Let S_x, S_y, S_z be the sets consisting of the orthogonal projections of the points of S onto the yz-plane, zx-plane, xy-plane, respectively. Prove that

$$|S|^2 \leq |S_x| \cdot |S_y| \cdot |S_z|$$

(Problem from the 1992 International Mathematical Olympiad).

Chapter 13

Recurrence

The method of recursion is important in exploring laws and solving problems from almost all branches of mathematics. The method is getting more and more attention with the widespread use of computers. The method of recursion for counting is as follows:

(1) denote by a_n the number count of objects that involves n;
(2) compute some initial values of $a_1, a_2, \ldots, a_k$;
(3) give recursive relations of a_{n+k} and $a_{n+k-1}, a_{n+k-2}, \ldots, a_n$, $n \geq 1$;
(4) solve (3) with initial values in (2).

Example 1. Find the number p_n of the permutations $(a_1, a_2, \ldots, a_n)$ of $(1, 2, \ldots, n)$, $n \geq 1$, with the property that there exists a unique $i \in \{1, 2, \ldots, n-1\}$, such that $a_i > a_{i+1}$.

Solution. It is easy to see that $p_1 = 0$, $p_2 = 1$.

For $n = m \geq 3$, in the case of $a_m = m$, the number of desired permutations is p_{m-1}.

If $a_i = m$, $1 \leq i \leq m-1$, select $i-1$ numbers from $(1, 2, \ldots, m-1)$ and put them in the increasing order $a_1, a_2, \ldots, a_{i-1}$, and the remaining numbers in the increasing order afterword. Then there are $\binom{m-1}{i-1}$ kinds of selections. Thus, the recursive relation is

$$p_m = p_{m-1} + \sum_{i=1}^{m-1} \binom{m-1}{i-1} = p_{m-1} + (1+1)^{m-1} - 1.$$

That is, $p_m - p_{m-1} = 2^{m-1} - 1$.

Summing up from $m = 2$ to $m = n$ yields

$$p_n = \sum_{m=2}^{n} (p_m - p_{m-1}) = \sum_{m=2}^{n} (2^{m-1} - 1) = 2^n - n - 1.$$

Example 2. Colour each edge of a convex n-gon ($n \geq 3$) with distinct edge lengths red, yellow, or blue, such that the neighbouring edges have different colours. How many distinct colourings are there?

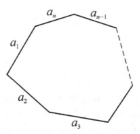

Figure 13.1

Solution. Suppose that there are p_n distinct colourings. It is easy to see that $p_3 = 6$.

If $n \geq 4$, first, there are three ways to colour edge a_1, so, there are two ways to colour edge a_2. Similarly, there are two ways to colour each edge $a_3, \ldots, a_{n-1}$. For edge a_n, we can colour it in two ways, but it over counts the case that a_1 and a_n are of the same colour. Thus, $p_n + p_{n-1} = 3 \times 2^{n-1}$. That is,

$$p_n - 2^n = -(p_{n-1} - 2^{n-1}).$$

Hence, $p_n - 2^n = (-1)^{n-3}(p_3 - 2^3) = (-1)^{n-2} \cdot 2$.
That is, $p_n = 2^n + (-1)^n \cdot 2$, $n \geq 3$.

Example 3. In a grid of $2 \times n$ squares, colour each square black or white, such that any two squares with a common side are not both black. How many distinct colourings are there?

Solution. Suppose that the number of colourings is a_n, in which the number of colourings with the nth column white is b_n. Thus, the number that the nth column has distinct colours is $a_n - b_n$.

For $n \geq 2$, if the $(n-1)$th column is white, then the nth column can be (from top to bottom) *white–white*, *black–white*, or *white–black*. If the

$(n-1)$th column has distinct colours, then the nth column has two ways of colouring. Thus,

$$a_n = 3b_{n-1} + 2(a_{n-1} - b_{n-1}) = 2a_{n-1} + b_{n-1}.$$

On the other hand, each colouring of $2 \times (n-1)$ squares corresponds to the colouring of $2 \times n$ squares with the nth column white. So, $b_n = a_{n-1}$. Thus, for $n \geq 3$, we have, $a_n = 2a_{n-1} + a_{n-2}$. And we can check that $a_1 = 3, a_2 = 7$. Hence,

$$a_n = \frac{1}{2}\left(1 + \sqrt{2}\right)^{n+1} + \left(1 - \sqrt{2}\right)^{n+1}.$$

Remark. A problem equivalent to Example 3 is as follows.

Let $\{a_n\}, \{b_n\}$ be two 0–1 sequences, such that

$$a_k a_{k+1} = b_k b_{k+1} = a_k b_k = 0 (k \in \mathbb{N}^+).$$

For given positive integer n, find the number of ways to assign values to $a_1, a_2, \ldots, a_n, b_1, b_2, \ldots, b_n$.

Example 4. In a permutation $(a_1, a_2, \ldots, a_n)$ of $(1, 2, \ldots, n)$, if $a_i \neq i$ for all $i = 1, 2, \ldots, n$, then we call such a permutation a derangement of $(1, 2, \ldots, n)$. Find the number of derangements D_n of $(1, 2, \ldots, n)$. (See also Example 3 of Chapter 5.)

Solution. It is easy to see that $D_1 = 0$, $D_2 = 1$.

For any derangement $(a_1, a_2, \ldots, a_n)$ of $(1, 2, \ldots, n)$ with $n \geq 3$, let $a_1 = k(k \neq 1)$, then,

(I) In case of $a_k = 1$, the number of derangements is D_{n-2}.
(II) In case of $a_k \neq 1$, treat the kth position as the first position and a_k as a_1, then $(a_2, \ldots, a_n)$ is a derangement of $(1, 2, \ldots, k-1, k+1, \ldots, n)$, and hence the number of derangements is D_{n-1}.

Since there are cases of $k = 2, 3, \ldots, n$, we have

$$D_n = (n-1)(D_{n-1} + D_{n-2}) \quad (n \geq 3). \tag{1}$$

Let $E_n = \frac{D_n}{n!}$, then (1) changes to $E_n = \left(1 - \frac{1}{n}\right)E_{n-1} + \frac{1}{n}E_{n-2}$, that is,

$$E_n - E_{n-1} = -\frac{1}{n} \cdot (E_{n-1} - E_{n-2}). \tag{2}$$

Use (2) repeatedly, and note that $E_1 = 0$, $E_2 = \frac{1}{2}$, for $n \geq 2$, we have

$$E_n - E_{n-1} = \left(-\frac{1}{n}\right) \cdot \left(-\frac{1}{n-1}\right) \cdot (E_{n-2} - E_{n-3}) = \cdots$$

$$= \left(-\frac{1}{n}\right) \cdot \left(-\frac{1}{n-1}\right) \cdots \left(-\frac{1}{3}\right) \cdot (E_2 - E_1) = \frac{(-1)^n}{n!}.$$

Hence, $E_n = \frac{(-1)^n}{n!} + \frac{(-1)^{n-1}}{(n-1)!} + \cdots + \frac{(-1)^2}{2!}$.

Therefore, $D_n = n! \left(1 - \frac{1}{1!} + \frac{1}{2!} - \cdots + (-1)^n \frac{1}{n!}\right)$.

Example 5. The code system of a new 'MO' lock is a regular n-gon, each vertex labelled a number 0 or 1 and coloured red or blue. It is known that for any two adjacent vertices, either their numbers or colours coincide. Find the number of all possible codes (in terms of n) (Problem from the 2010 China High School Mathematical League).

Solution. Let a_n be the number of different key codes. We shall prove that

$$a_n = 3^n + 2 + (-1)^n.$$

For convenience, replace each vertex configuration by an integer $k, 0 \leq k \leq 3$, as illustrated in the following table:

k	Red	Blue
0	0	3
1	1	2

Each key code corresponding to a sequence $\{A_1, A_2, \ldots, A_n\}$, $0 \leq A_i \leq 3$. Clearly, neighbouring vertices must have the same or consecutive numbers (0 and 3 are considered consecutive). Define $B_i \equiv A_i - A_{i-1} \pmod 4$, and $|B_i| \leq 1$ for all $1 \leq i \leq n$ ($B_1 = A_1 - A_n$).

On the other hand, if we choose $0 \leq A_1 \leq 3$ and $|B_i| \leq 1$ for $2 \leq i \leq n$, then they determine a valid key code if $|B_1| = |-\sum_{i=2}^n B_i| \leq 1 \pmod 4$. Therefore, each code, or sequence $\{A_1, A_2, \ldots, A_n\}$ is uniquely determined by A_1 and $B_1, B_2, \ldots, B_n$ satisfying

$$S = \sum_{i=1}^n B_i \equiv 0 \pmod 4.$$

Furthermore, we infer that a_n is the number of sequences

$$\{A_1, B_1, B_2, \ldots, B_n\}, \quad 0 \le A_1 \le 3, |B_i| \le 1, \ S \equiv 0 \ (\text{mod } 4). \qquad (1)$$

Let b_n, c_n, d_n be the number of sequences similar to a_n but $S \equiv 1, 2, 3$ (mod 4), respectively. It is easy to see that $a_n + b_n + c_n + d_n = 4 \cdot 3^n$ for all $n \ge 1$.

For $n \ge 2$, each sequence in (1) can be obtained by adding $B_n = 0, -1, 1$ to a sequence $\{A_1, B_1, B_2, \ldots, B_{n-1}\}$, $0 \le A_1 \le 3, |B_i| \le 1$, $S \equiv 0, 1$, or 3 (mod 4), and the number of these sequence are $a_{n-1}, b_{n-1}, d_{n-1}$, respectively, but (1) cannot be obtained from a sequence $\{A_1, B_1, B_2, \ldots, B_{n-1}\}$ with $S \equiv 2$ (mod 4), namely from c_{n-1}. Thus,

$$a_n = a_{n-1} + b_{n-1} + d_{n-1} = 4 \cdot 3^{n-1} - c_{n-1}.$$

Similarly,

$$c_{n-1} = b_{n-2} + c_{n-2} + d_{n-2} = 4 \cdot 3^{n-2} - a_{n-2}.$$

From the above two equations, we have the recursive formula

$$a_n = 8 \cdot 3^{n-2} + a_{n-2} \quad n = 3, 4, \ldots.$$

It is easy to see $a_1 = 4$, $a_2 = 12$, and hence: (i) if n is odd,

$$a_n = 8(3^{n-2} + 3^{n-4} + \cdots + 3^1) + a_1 = 3^n + 1;$$

(ii) if n is even,

$$a_n = 8(3^{n-2} + 3^{n-4} + \cdots + 3^2) + a_2 = 3^n + 3.$$

This completes the proof.

The following example shows that the applications of the idea of recursion are not simply by means of recursion, but are flexible and have various forms.

Example 6. Let two non-negative real sequences $\{a_n\}, \{b_n\}$ satisfy: $a_i + b_j \le a_i b_j$ for any $i, j \in \mathbb{N}^+$, $|i - j| \in \{2011, 2012\}$. Prove that all terms of $\{a_n\}, \{b_n\}$ are either zeros or greater than 1.

Solution. If for any $n \in \mathbb{N}^+$, $a_n, b_n > 1$, then there is nothing to prove. In the following, suppose a term in $\{a_n\}, \{b_n\}$, say a_k, is less than 1.

Since inequality $a_i + b_j \le a_i b_j$ is equivalent to $(1 - a_i)(1 - b_j) \ge 1$, if $0 \le a_i < 1$, then

$$1 \ge 1 - b_j \ge \frac{1}{1 - a_i} \ge 1.$$

We see that all equalities hold, hence $a_i = b_j = 0$.

Since $0 \leq a_k < 1$, we let $(i, j) = (k, k + 2011), (k, k + 2012)$ and obtain

$$a_k = b_{k+2011} = b_{k+2012} = 0.$$

Then let $(i, j) = (k - 1, k + 2011), (k + 1, k + 2012)$ (if $k = 1$, omit the former), we have

$$a_{k-1} = a_{k+1} = 0.$$

In the same manner, we have

$$a_{k-2} = a_{k-3} = \cdots = a_1 = 0, \quad a_{k+2} = a_{k+3} = \cdots = 0.$$

That is, each term in $\{a_n\}$ is 0. Moreover, there is a 0 in $\{b_n\}$, by the same argument, each term in $\{b_n\}$ is 0. This completes the proof.

Remark. The conclusion is valid if the condition in Example 6 is relaxed to a more general case: for any $i, j \in \mathbb{N}^+$, $|i - j| \in \{p, q\}$, $(p, q) = 1$, and positive integers p and q are not both odd.

Example 7. For odd integer $n > 1$, prove that the real zero α of $P(x) = (x - 1)^n - x^2$ is unique and $\alpha \in (b, c) = \left(1 + \exp\left(\frac{\log 4}{n-1}\right), 1 + \exp\left(\frac{\log 4}{n-2}\right)\right)$, where $\exp(x) = e^x$ is the exponential function.

Solution. First, let $x = 2 + y$, $y > -1$, then $P(x) = 0$ is equivalent to

$$\frac{(1 + y)^{n/2}}{2(1 + y/2)} = 1.$$

Take the logarithm on both sides of the above equation, obtain the equivalent equation $f(y) = \frac{n}{2} \log(1 + y) - \log(1 + y/2) - \log 2 = 0$.

And we see that if $f(y) > 0$, then $P(x) > 0$, if $f(y) < 0$ then $P(x) < 0$.

By $f'(y) = \frac{n}{2(1+y)} - \frac{1}{2+y} = \frac{2(n-1)+(n-2)y}{2(1+y)(2+y)} > 0$.

So, $f(y)$ is strictly increasing on $y > -1$.

Rewrite

$$f(y) = \frac{n-1}{2} \log(1 + y) - \log 2 - \frac{1}{2} \log\left(1 + \frac{y^2}{4(1+y)}\right).$$

Then we see that if $x = b$, then $y = \exp\left(\frac{2\log 2}{n-1}\right) - 1 > 0$,

$$f(y) = -\frac{1}{2} \log\left(1 + \frac{y^2}{4(1+y)}\right) < 0,$$

thus, $P(b) < 0$.

Rewrite

$$f(y) = \frac{n-2}{2} \log(1+y) - \log 2 + \log\left(1 + \frac{y}{2+y}\right).$$

Then we see if $x = c$, then $y = \exp\left(\frac{2\log 2}{n-2}\right) - 1 > 0$,

$$f(y) = \log\left(1 + \frac{y}{2+y}\right) > 0,$$

so, $P(c) > 0$.

Summing up, there is a unique real zero α of $P(x)$ on interval $x > 1$, and $\alpha \in (b, c)$.

Next, we should prove that there are no zeros of $P(x)$ on $x \leq 1$.

In fact, if $x \leq 0$, $P(x) = (x-1)^n - x^2 \leq -1 < 0$.

If $0 < x \leq 1$, $P(x) < (1-1)^n - 0^2 = 0$.

This, completes the proof.

Remark. If we want to compute the value of real zero α of $P(x)$, we can use the method of recurrence $a_{k+1} = 1 + a_k^{2/n}$, $k \in \mathbb{N}$ with $a_0 = 0$.

We prove that $\{a_k\}$ is a strictly increasing sequence by induction.

(I) $a_{k+1} - a_k = 1 + a_k^{2/n} - a_k > 0$ is true for $k = 0$.

(II) Suppose that for $k \geq 0$, $1 + a_k^{2/n} > a_k$, then

$$a_{k+2} - a_{k+1} = 1 + a_{k+1}^{2/n} - a_{k+1}$$

$$= 1 + (1 + a_k^{2/n})^{2/n} - (1 + a_k^{2/n})$$

$$> 1 + a_k^{2/n} - (1 + a_k^{2/n}) = 0.$$

This completes the induction.

For $a_k < x \leq a_{k+1}$, we see that $P(x) < (a_{k+1} - 1)^n - a_k^2 = 0$.

Since $\{a_k\}$ has an upper bound c, so $a = \lim_{k \to +\infty} a_k > 0$ exists and $(a-1)^n - a^2 = 0$, that is, a is a real root of $P(x) = 0$.

Example 8. There are three piles of matches on a table. The number of matches of one pile is the sum of that of the other two piles. Two players play the game in turn with the rule that one player takes away any one pile of matches, and divides any one pile left into two nonempty piles. The one that cannot do this loses. Prove or disprove that the first player has a winning strategy.

Solution. We prove the first player has a winning strategy. For the current state, let the numbers of matches of three piles be $2^{n_1}a_1, 2^{n_2}a_2, 2^{n_3}a_3$, respectively (where $n_1, n_2, n_3 \in \mathbb{N}$, a_1, a_2, a_3 are all odd numbers). Define the state by L if $n_1 = n_2 = n_3$, and by W otherwise.

Firstly, we see that the initial state is W, for if it is L, then we may suppose that $2^{n_1}a_1 + 2^{n_2}a_2 = 2^{n_3}a_3$. As $n_1 = n_2 = n_3$, $a_1 + a_2 = a_3$, which contradicts the fact that a_1, a_2, a_3 are all odd.

Secondly, the first player can change state W to state L. Suppose that $n_1 \leq n_2 \leq n_3$, by the definition of state W, we have $n_1 < n_3$. Player A takes away the second pile and divides the third pile into two piles of 2^{n_1}, $2^{n_1}(2^{n_3-n_1}a_3 - 1)$ matches, which is state L.

Thirdly, if the second player can go, then he can only change state L into state W. In fact, let the state L be $2^n a_1, 2^n a_2, 2^n a_3 (n \in \mathbb{N}, a_1, a_2, a_3$ are all odd numbers). Suppose on the contrary that the result is a state L, and the third pile remains the same, then the resulting state is $2^n b_1, 2^n b_2$, $2^n a_3 (b_1, b_2, a_3$ are all odd numbers), where $b_1 + b_2 = a_1$ or $= a_2$, which contradicts the fact that a_1, a_2, b_1, b_2 are all odd numbers.

Since the game must end in finite steps, the first player has the winning strategy.

Remark. Two-player game problems often appear in mathematics competitions: the game is finite and zero-sum, and the two sides of the game operate in turn according to the rules until the winner or the loser is determined. When both sides take the best approach, the winner (and sometimes the undefeated) must be identified. The general steps to solve such game problems are as follows:

1. Work out the game results for small numbers to make reasonable guesses on the *win* state (that is, the situation in which appropriate actions can be taken to win eventually) and the *negative state* (that is, no matter what actions are taken, the other side can always take appropriate actions to win);
2. Verify that a *negative state* can indeed be obtained from each *win* state when acting appropriately;
3. Verify that no matter what a player does from each *negative state*, only the *win* state will be obtained.

So, you can determine the outcome for each initial state.

This problem is a revised form of a question from the 1994 All-Russian Mathematical Olympiad. Further, we can consider the problem: initially there are four piles of matches, where one pile has the sum of another two piles. If the rule remains the same, does the first player have a winning strategy? (Problem 8 of Exercises is also a relevant problem)

Exercises

1. There are 12 people standing in a line in a bank. When the desk closes, the people form a new line at a newly opened desk. In how many ways can they do this in such a way that none of the 12 people changes his/her position in the line by more than one? (Problem from the 2005 Swedish Mathematical Olympiad.)

2. Two persons throw a die in a game with the rule that if a person throws one dot, then throws again until it is more than one dot, then they should let the other person throw. Let the probability of the first player throwing at the nth throw be A_n. find A_n.

3. Consider n-digit numbers whose digits are all from $\{1, 2, 3, 4, 5\}$. If the number of n-digit numbers that contain 5 with no 3 ahead of 5 is $f(n)$, find $f(n)$.

4. Find the coefficient of x^2 in the following expression after expansion in increasing order of x:

$$((\cdots(((x-2)^2 - 2)^2 - 2)^2 - \cdots - 2)^2 - 2)^2, \quad (n \text{ pairs of parentheses}).$$

5. In a standard deck of 52 playing cards, there are four suits (diamonds, clubs, hearts, spades), each with 13 ranks $2, 3, \ldots, 10, J, Q, K, A$. Two consecutive cards of the same suit are called *straight flush*. (A and 2 are consecutive.) Find the number of ways of taking 13 cards with distinct ranks and 2 cards without *straight flush*.

6. set $S \subset \mathbb{N}^+$ to satisfy that if $a, b \in S$, $a < b$, then $ab + 3 \in S$. Prove that there are infinitely many multiples of 7 in S.

7. Given an $m \times n$ grid with squares coloured either black or white, we say that a black square in the grid is *stranded* if there is some square to its left in the same row that is white and there is some square above it in the same column that is black. Find a closed formula for the number of $2 \times n$ grids with no *stranded* black squares. The figure shows a 4×5 grid without *stranded* black squares (Problem from the 2009 Canadian mathematical Olympiad).

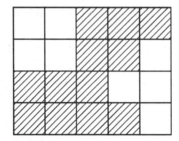

8. There are four piles of matches on the table for a game of two players. They take away matches in turn following the rule of taking away any pile of the matches and dividing one remaining pile into two nonempty piles. The one that cannot follow the rule loses. At the beginning, there are 3 piles of 2 matches and 1 pile of n matches. Find all the possible values of n, such that the first player has a winning strategy.

Chapter 14

Colouring

Some mathematical problems, such as operational problems and logical reasoning problems, cannot be solved by ordinary mathematical methods. There are also practical problems, that deal with a certain state or property of things, which themselves have nothing to do with quantity, and should not be solved by ordinary mathematical methods. These unconventional mathematical problems require specific methods, such as colouring, and assignment which is described in the next chapter.

Colouring is a visualization method to classify the objects. By means of colouring, the relatively abstract combinatorics problem can be transformed into a specific colouring problem, and this helps us to observe and analyze the relationship between objects, and then to solve the original problem through the processing of colouring patterns. Common colouring methods include point colouring, edge colouring, grid colouring, and region colouring.

We illustrate the application techniques of colouring by some examples.

Example 1. Prove that among any six people, there are at least three people who know each other or do not know each other.

Solution. Regard six people as six vertices $v_1, v_2, \ldots, v_6$. If two people know each other, then link the corresponding vertices with a red edge, otherwise a blue edge. The problem now is to prove there is a monochromatic triangle (that is, a triangle with all sides in a colour).

Consider five edges (v_i, v_j), $j \neq i$, incident with v_i. By the drawer principle, there are at least three edges (v_i, v_j), (v_i, v_k), (v_i, v_l) in a colour, say, red. If $\triangle v_j v_k v_l$ is blue, then the conclusion is true; otherwise there is a red edge, say, (v_j, v_k), then $\triangle v_i v_j v_k$ is red.

Remark. This is a classic example of a colouring problem that has many variants. By colouring the edges, we turn the problem into the famous Ramsey proposition in graph theory (c.f. Chapter 20): in a two-colouring complete graph K^6 there exists at least one monochromatic triangle.

In addition, since each edge is coloured, we can give another proof as follows by counting monochromatic angles (that is, the angles composed of two edges in a colour):

Let the number of monochromatic triangles in K^6 be n. There are totally 20 triangles, so, there are totally $(20 - n)$ heterochromatic triangles (triangle not in a colour).

On the one hand, each monochromatic triangle has three monochromatic angles, and each heterochromatic triangle has one monochromatic angle, so the total number of monochromatic angles is $S = 3n + (20 - n) = 2n + 20$. On the other hand, if vertex v_i is incident with r_i red edges, then it is incident with $5 - r_i$ blue edges, so the number of the monochromatic angles at the vertex v_i is

$$S_i = \binom{r_i}{2} + \binom{5 - r_i}{2} = (r_i - 2)(r_i - 3) + 4.$$

Thus, $2n + 20 = S = \sum_{i=1}^{6} S_i = 24 + \sum_{i=1}^{6}(r_i - 2)(r_i - 3) \geq 24$. That is, $n \geq 2$. Equality holds if and only if each $r_i = 2$ or 3. For example, let $\triangle v_1 v_2 v_3$ and $\triangle v_4 v_5 v_6$ be red, and the other nine edges be blue. Another example is to let $\triangle v_1 v_2 v_3$ be red and $\triangle v_1 v_5 v_6$ be blue, and edges $(v_1, v_4), (v_2, v_6), (v_3, v_5), (v_4, v_5), (v_4, v_6)$ be red, and edges $(v_2, v_4), (v_2, v_5), (v_3, v_4), (v_3, v_6)$ be blue.

We have just proved a stronger conclusion: there are at least two monochromatic triangles in a two-colour complete graph K^6.

The above two methods can be regarded as typical methods to prove the colouring problem of graphs.

Example 2. How many cargoes of size $1 \times 2 \times 4$ can a box of capacity $6 \times 6 \times 6$ contain?

Solution. Divide the box space into 27 medium cubes of size $2 \times 2 \times 2$ with alternating colourings in black and white as in Fig. 14.1. Altogether 14 black medium cubes and 13 white medium cubes. Then divide each medium cube into eight small cubes. There are totally 216 small cubes, 112 black and 104 white.

Let k be the maximum capacity of cargoes. Obviously, $k \leq \frac{216}{8} = 27$.

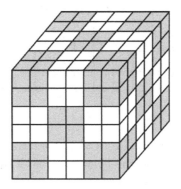

Figure 14.1

If $k = 27$, then the box must be filled without space. Each cargo takes up the space of four small black cubes and four small white cubes. But the total numbers of black and white small cubes are different, a contradiction. Thus, $k \leq 26$.

On the other hand, there are 13 pairs of neighbouring medium cubes, one black and one white, each pair can contain two cargoes, therefore $k = 26$.

Example 3. Show that on a chessboard of 4×8 squares, a *horse* cannot start from one square, visit every square on the board exactly once, and return to its starting square. (A *jump* step of the *horse* is first moving two squares vertically or horizontally, then moving one square to its left or right.)

Solution. Suppose on the contrary, the *horse* can visit each square once and return. We may let the starting square A be on the upper left corner.

First, colour the squares alternating in black and white as Fig. 14.2. (A is white.) At each step, the *horse jumps* between two squares with different colours. So, each black square is visited at odd steps, and each white square is visited at even steps.

On the other hand, colour the first and the fourth rows white and the second and the third rows black. We see that the *horse* at a white square

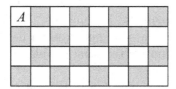

Figure 14.2

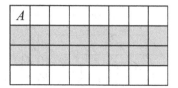

Figure 14.3

can only *jump* to a black square. Thus, there are 16 *jumps* from white squares to black squares, and the *horse* on a black square must jump to a white square. So, each black square is visited at odd steps and every white square is visited at even steps. Obviously the two sets of black squares in Figs. 14.2 and 14.3 are different. A contradiction.

Remark. This example uses two colouring methods. The first colouring method is often called 'natural colouring', which is often related to pairing and parity. The second colouring method shows the flexibility of solving specific problems and can be regarded as an unconventional colouring method. In this problem, we fully consider the characteristics of the chessboard and the *horse*, using the conventional method and the unconventional method, each method reveals an invariant rule in the process of the *horse* traversing the chessboard, and then the mutual contrast leads to contradiction, so that the proposition is proved.

Example 4. Two players A and B are playing a game on an infinite chessbard. They take turns to place pawns on empty squares. A goes first and places a black pawn, and B, white. The player who places five consecutive pawns of his colour vertically or horizontally wins. Does player A have a winning strategy?

Solution. The answer is negative. We show that player B has a strategy to prevent A winning.

Divide the chessboard into 2×2 big squares, and colour the big squares alternately black and white as in Fig. 14.4.

If player A puts a pawn in a black big square, then player B puts a pawn at the same big square on the same row. If player A puts a pawn in a white big square, then player B puts a pawn in the same white square on the same column. Adopting this strategy, player B always has a unique move whatever player A does.

For any consecutive five powns in a vertical or horizontal line, there must be an adjacent pair in a black big square and an adjacent pair in a

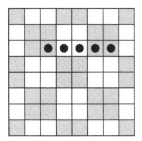

Figure 14.4

white big square. However, based on player B's strategy, one adjacent pair is not in the same colour. Therefore, player A cannot win.

Example 5. On a 10×10 chessboard, some $4n$ unit squares are chosen to form a region R. This region R can be tiled by n 2×2 squares. This region R can also be tiled by a combination of n tetrominoes '⊞' or '⊞'. (with rotations allowed) Determine the minimum of n (Problem from the 2009 China Girls' Mathematical Olympiad).

Solution. First, we show that n is even. Part of the chessboard shown as Fig. 14.5.

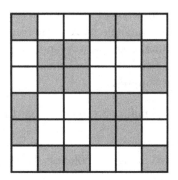

Figure 14.5

Call the shape ⊞ A and the shapes ⊞ and ⊞ B (with rotations allowed). No matter which four squares A covers, the number of black squares is even, while the number of black squares B covers is odd. Therefore, n must be even.

If $n = 2$, the region tiled by $2A$ can only be as the following, which cannot be tiled by $2B$ (Fig. 14.6).

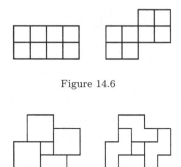

Figure 14.6

Figure 14.7

Therefore, $n \geq 4$. The following regions are tiled by 4 A and 4 B. So, $n = 4$ (Fig. 14.7).

Example 6. On an infinite chessboard, each square is allowed to hold a chess piece. Two pieces are '*adjacent*' if they are on squares that have at least a common vertex. At the beginning, 81 chess pieces are placed as a 9×9 square. The playing rule is the following:

(I) If a piece X has an adjacent piece Y horizontally or vertically, put a piece on the square symmetric to X with respect to Y if there is no piece, at the same time, take away X and Y.

(II) If a piece X has no adjacent pieces, then take it out, and put pieces on its eight adjacent squares.

Can only one chess piece remain on the chessboard in finite steps?

Solution. The answer is negative.

Colour the chessboard in three colours A, B, and C as shown in Fig. 14.8.

For rule (I), the number of pieces on the chessboard of any colour decreases or increases by 1. For rule (II), the number of pieces of a colour increases by 1, and that of every other colour increases by 3. Thus, the number of pieces on the chessboard of each colour changes its parity simultaneously.

At the beginning, the number of pieces of each colour is odd (27,27,27). If there is only one piece left on the chessboard, then the numbers are odd, even, and even (1,0,0). A contradiction.

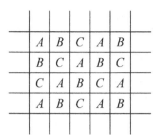

Figure 14.8

Remark. (a) If 9×9 is changed into $n \times n$ in Example 6, one can remove all but one piece if n is not a multiple of 3, one can remove all but one piece in finite steps (the steps can be constructed by induction).

(b) The proof is also valid if in rule (I) the adjacent relation of pieces X and Y also includes upper left and lower right.

(c) If in rule (I), the adjacent relation of pieces X and Y also includes lower left and upper right, then colouring the chessboard by its mirror image, the proof is also valid by symmetric property.

But if relaxing the rule (I) to both (b) and (c), then the answer is positive, which is left to the reader to prove.

Example 7. There is a non-self-intersecting closed polyline on a 15×15 grid, which is composed of several line segments connecting the centres of adjacent small squares (two small squares with a common side are called adjacent), and it is symmetric with respect to a diagonal of the grid. Prove that the length of this closed polyline is no more than 200.

Solution. Let A be an intersection point of the polyline and the diagonal. Moving along the polyline starting from point A, let B be the first point that meets the diagonal again. Moving along the polyline from point A in the opposite direction, B is still the first point met by the symmetricity. Since the polyline is non-self-intersecting, A and B are the only intersect points of the polyline and the diagonal.

Now we colour the grid in black and white alternately such that the squares on the diagonal are black. Black and white squares are interlacing along the polyline. Thus, the numbers of the black and white squares the polyline passes are equal. Since the polyline misses 13 black squares on the diagonal and the number of black squares is one more than the white ones, the polyline misses at least 12 white squares. Hence, the length of the polyline is no more than $15^2 - 13 - 12 = 200$.

Example 8. $3k$ points are marked on the circumference. They divide it onto $3k$ arcs. Some k of them have length 1, other k of them have length 2, and the rest k of them have length 3. Prove some two of the marked points are the ends of one diameter (Problem from the 1982 All-Soviet Union Mathematical Olympiad).

Solution. Colour $3k$ points in red, and colour the midpoint of each arc of length 2 and two trisection points of each arc of length 3 in blue. Obviously, there are $3k$ blue points. Thus, the problem is to prove some two red points are the ends of a diameter.

Suppose the conclusion does not hold: the antipodal point of every red point is blue. Since the number of red points is the same as that of blue points, the antipodal point of every blue point is red. Take any arc (denoted by AC) with length 2, AC is red, and the midpoint B' of AC is blue, then the antipodal point B of B' is red.

Consider arc AB of length $3k - 1$, on which the number of arcs of length i is $n_i (i = 1, 2, 3)$. Then $n_1 + 2n_2 + 3n_3 = 3k - 1$. Since the endpoints of an arc of length 1 are red, their antipodal points are blue and are trisection points of an arc of length 3 on arc BC. That is a one-to-one correspondence. So, the number of arcs of length 3 on BC is n_1. On the other hand, there are k arcs of length 3 on AB and BC together. Hence, $n_3 + n_1 = k$. Consequently,

$$2n_2 + 2n_3 = (n_1 + 2n_2 + 3n_3) - (n_3 + n_1) = (3k - 1) - k = 2k - 1.$$

In the above equation, the left-hand side is even and the right-hand side is odd, a contradiction.

Exercises

1. Can a grid of 8×8 squares with two squares on the opposite corners cut out be covered by 31 dominoes of 2×1 squares?

2. Divide a cubic house into 27 small rooms of equal size. There is a door between each two neighbouring rooms. A beetle starts a journey from the center room. It can go to any neighbouring room. Can the beetle visit each room once in all?

3. Divide a equilateral triangle into n^2 small equilateral triangles of equal size. Label m of these small equilateral triangles with numbers

$1, 2, \ldots, m$, such that any two small equilateral triangles labeled with neighbouring numbers have a common side. Prove that $m \leq n^2 - n + 1$.

4. A convex n-gon is divided by disjoint diagonals into triangles, such that there is an odd number of triangles at each vertex of the n-gon. Prove that $3 | n$.

5. If in any 3 persons of a group of 9 people, there are 2 persons knowing each other, prove that there are 4 persons knowing each other.

6. There are 6 points on a plane, any 3 points of which are the vertices of a triangle with unequal sides. Prove that there are two triangles T_1 and T_2, such that the shortest side of T_1 is the longest side of a triangle T_3, and the longest side of T_2 is the shortest side of a triangle T_4 (T_3 and T_4 may be the same triangle).

7. Faces of a $9 \times 9 \times 9$ cube are partitioned onto unit squares. The surface of the cube is pasted over by 243 strips of 2×1 without overlapping. Prove that the number of bent strips is odd (Problem from the 2007 All-Russian Mathematical Olympiad).

8. Write any positive integer at each square of an 8×8 chessboard. Define an operation as follows. Take any *sub chessboard* of 2×2, 3×3, or 4×4, and add 1 to all numbers on the *sub chessboard*. Is it possible aways to make all numbers on the chessboard even after finite operations?

9. What is the maximum number of non-overlapping 2×1 dominoes that can be placed on an 8×9 checkerboard if six of them are placed as shown in Fig. 14.9? Each domino must be placed horizontally or vertically so as to cover two adjacent squares of the board (Problem from the 2007 Canadian Mathematical Olympiad).

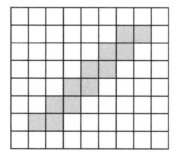

Figure 14.9

Chapter 15

Assignment Methods

Assignment is a method to solve problems that have nothing to do with quantity by cleverly assigning some appropriate values to certain objects, making them numerical, and solving the problem by the discussion of divisibility, parity, or sign, etc.

Many combinatorial problems and non-traditional number theory problems are often solved by assignment. Note that in colouring, things are classified by colours, and that assignment can replace 'colour' with 'number', so in general, problems that can be expressed and solved by colouring can also be dealt with by assignment (although colouring has its advantages in terms of visualization, etc.). The assignment method makes the problem numerical, and further involves the numerical values in the operation and the deduction, therefore, it has its unique display space. Common ways of assigning values are: assigning values to points, line segments, regions, and other objects.

Example 1. There are n points $A_1, A_2, \ldots, A_n$ located on a line in order as shown in Fig. 15.1. Each point is coloured red or blue. We call the segment $A_i A_{i+1}$ $(1 \leq i \leq n-1)$ a *standard* one, if its endpoints are in different colours. If A_1 and A_n are in different colours, show that the number of *standard* segments in $A_i A_{i+1}$ $(i = 1, 2, \ldots, n-1)$ is odd.

$$A_1 \quad A_2 \quad A_3 \quad A_4 \qquad\qquad A_{n-1} \quad A_n$$

Figure 15.1

Solution. Assign each red point A_i the number $a_i = 1$ and blue point A_j the number $a_j = -1$. So, for a *standard* segment, $A_i A_{i+1}$, $a_i a_{i+1} = -1$. And for a nonstandard segment, $A_i A_{i+1}$, $a_i a_{i+1} = 1$.

Then,

$$(a_1 a_2)(a_2 a_3) \cdots (a_{n-1} a_n) = a_1 a_2^2 a_3^2 \cdots a_{n-1}^2 a_n = a_1 a_n = -1.$$

Hence, the number of *standard* segments is odd.

Example 2. There are n students sitting around a round table. A red flower is put between every two neighbouring students of the same sex, and a blue flower is put between every two neighbouring students of the opposite sex. If the numbers of red and blue flowers are equal, show that n must be a multiple of 4.

Solution. Since the numbers of red and blue flowers are equal, n must be even. Let $n = 2m$. Label n $(n > 2)$ students with $1, 2, \ldots, n$ in order. Assign the ith $(1 \le i \le n)$ student a number $x_i = 1$ for male or $x_i = -1$ for female.

Thus, there are m of 1s and m of (-1)s in the n products $x_i x_{i+1}$ $(1 \le i \le n)$, $(x_{n+1} = x_1)$.

Hence, $(-1)^m = \prod_{i=1}^{n} x_i x_{i+1} = (x_1 x_2 \cdots x_n)^2 = 1$. That is, m is even. Therefore, $n = 2m$ is a multiple of 4.

Example 3. Put 2004 numbers $1, 2, 3, \ldots, 2004$ on a circle at will. After checking the parity of every three neighbouring numbers, one finds 600 groups with three odd numbers, 500 groups with exactly two odd numbers. How many groups are there with just 1 odd number? How many groups are there without odd numbers? (from the 2004 Mathematical Competition of High Schools for Grade 2 in Shanghai, China)

Solution. Let the number of groups with exactly one odd number be x, then the number of groups without odd numbers is

$$2004 - 600 - 500 - x = 904 - x.$$

Denote the numbers on the circle by $x_1, x_2, \ldots, x_{2004}$, and let

$$y_i = \begin{cases} -1, & \text{if } x_i \text{ is odd,} \\ 1, & \text{if } x_i \text{ is even.} \end{cases}$$

Then $y_1 + y_2 + \cdots + y_{2004} = 0$. Also,

$$A_i = y_i + y_{i+1} + y_{i+2}$$

$$= \begin{cases} -3, & \text{if } x_i, x_{i+1}, x_{i+2} \text{ are all odd,} \\ -1, & \text{if } x_i, x_{i+1}, x_{i+2} \text{ are 2 odd and 1 even,} \\ 1, & \text{if } x_i, x_{i+1}, x_{i+2} \text{ are 2 even and 1 odd,} \\ 3, & \text{if } x_i, x_{i+1}, x_{i+2} \text{ are all even,} \end{cases} \quad (x_{2004+i} = x_i \ i = 1, 2)$$

Hence,

$$0 = 3(y_1 + y_2 + \cdots + y_{2004}) = A_1 + A_2 + \cdots + A_{2004}$$
$$= -3 \times 600 - 500 + x + 3(904 - x),$$

which yields $x = 206$ and $904 - x = 698$.

That is, there are 206 groups with just 1 odd and 698 groups without any odd number.

Example 4. Cover a chessboard of size 5×7 by several pieces of paper. The paper can overlap, but the edge of the paper must fit the grids of the chessboard. By using one size of paper (pieces) as follows, can you cover the chessboard with the same number of layers of papers on each square? (Paper can be turned over)

(a) size 1×3;
(b) size of 'L' tromino, that is, paper 2×2 with a square cut at corner;
(c) size of 'P' pantomino that is, paper 2×3 with a square cut at corner.

Solution. (a) Assign each square a number as shown in Fig. 15.2. You can see that the sum of any three numbers covered by paper size 1×3 is 0. Therefore, no matter how the paper covers the board and how many sheets of paper are used, the sum of the numbers covered is 0. But the total sum of the numbers on the chessboard is 1, and it is n for n layers.

Thus, the answer for (a) is negative.

(b) Assign numbers as Fig. 15.3. It is easy to see that the sum of numbers covered by an 'L' tromino is no more than 0. But the total sum of the numbers on the chessboard is 1. So, the answer for (b) is negative.

(c) Assign numbers as Fig. 15.4. It is easy to see that the sum of numbers covered by a 'P' pentomino is no less than 0. But the total sum of the numbers on the chessboard is -5. Thus, the answer for (c) is negative.

2	−1	−1	2	−1	−1	2
−1	−1	2	−1	−1	2	−1
−1	2	−1	−1	2	−1	−1
2	−1	−1	2	−1	−1	2
−1	−1	2	−1	−1	2	−1

Figure 15.2

2	−1	2	−1	2	−1	2
−1	−1	−1	−1	−1	−1	−1
2	−1	2	−1	2	−1	2
−1	−1	−1	−1	−1	−1	−1
2	−1	2	−1	2	−1	2

Figure 15.3

−1	−1	−1	−1	−1	−1	−1
−1	4	−1	4	−1	4	−1
−1	−1	−1	−1	−1	−1	−1
−1	4	−1	4	−1	4	−1
−1	−1	−1	−1	−1	−1	−1

Figure 15.4

Example 5. Figure 15.5 is a grid infinitely towards right and downward. At first, on the grid, a stone is at the upper left square A. Then the rule is: if at some step, there is a stone at P, and there are no stones at its right and downward neighbouring squares Q and R, one can remove the stone at P and put stones at Q and R. Prove that there are/is always stone(s) at the upper left 3×3 squares at any step.

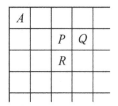

Figure 15.5

Solution. Assign the square at ith row and jth column the number $\left(\frac{1}{2}\right)^{i+j}$, $i, j \in \mathbb{N}^+$. Since

$$\left(\frac{1}{2}\right)^{i+(j+1)} + \left(\frac{1}{2}\right)^{(i+1)+j} = \left(\frac{1}{2}\right)^{i+j},$$

the sum of numbers of the occupied squares, denoted by S, is unchanged at each step. At first $S = \frac{1}{4}$.

Assume, on the contrary, there were no stones on the upper left 3×3 squares at some step, then

$$S \le \sum_{i=1}^{\infty} \sum_{j=1}^{\infty} \left(\frac{1}{2}\right)^{i+j} - \sum_{i=1}^{3} \sum_{j=1}^{3} \left(\frac{1}{2}\right)^{i+j}$$

$$= \sum_{i=1}^{\infty} \left(\frac{1}{2}\right)^{i} \cdot \sum_{j=1}^{\infty} \left(\frac{1}{2}\right)^{j} - \sum_{i=1}^{3} \left(\frac{1}{2}\right)^{i} \cdot \sum_{j=1}^{3} \left(\frac{1}{2}\right)^{j}$$

$$= 1 - \left(\frac{7}{8}\right)^2 = \frac{15}{64} < \frac{1}{4}.$$

A contradiction.

Example 6. We have k switches arranged in a row, and each switch points up, down, left, or right. Whenever three successive switches all point in different directions, all three may be simultaneously turned so as to point in the fourth direction. Prove that this operation cannot be repeated infinitely many times (Problem from the 2006 Bay Area Mathematical Olympiad).

Solution. Label these k switches with $1, 2, \ldots, k$ in order. Assign the nth ($1 \le n \le k$) switch the number $f(n)$ according to the following rule:
   If $n = 1$, $f(n) = 1$.

If $n \geq 2$, and the direction of the $(n - 1)$th switch and direction of the nth switch are the same, then $f(n) = n$; otherwise, $f(n) = 1$.

Define *height* $h = f(1)f(2) \cdots f(k)$, we show that h is increasing after each operation.

For any n, suppose that the nth, $(n + 1)$th, and the $(n + 2)$th switches point to three different directions, then $f(n) \leq n$, $f(n + 3) \leq n + 3$, $f(n + 1) = f(n + 2) = 1$.

Thus,

$$f(n)f(n + 1)f(n + 2)f(n + 3) \leq n(n + 3).$$

After the operation, $f'(n) \geq 1$, $f'(n+3) \geq 1$, $f'(n+1) = n+1$, $f'(n+2) = n + 2$.

Thus,

$$f'(n)f'(n + 1)f'(n + 2)f'(n + 3) \geq (n + 1)(n + 2) > n(n + 3).$$

Hence,

$$f'(n)f'(n + 1)f'(n + 2)f'(n + 3) > f(n)f(n + 1)f(n + 2)f(n + 3).$$

Since $f'(i) = f(i)$, for $1 \leq i \leq n - 1$ and $n + 4 \leq i \leq k$, we have $h < h'$.

Since $h \in \mathbb{N}^+$ and $h \leq 1 \times 2 \times \cdots \times k$, h can increase only finite number of times. And the result follows.

Remark. (a) We can also define other *height* functions. Assign the nth switch the value $f(n)$: If the nth and the $(n + 1)$th switches have different directions, then $f(n) = \sqrt{n}$, otherwise $f(n) = 0$. The *height* $h = \sum_{n=1}^{k} f(n)$. We can prove $h > 0$ is decreasing after the operation, hence h can only decrease finitely many times which completes the proof.

(b) Readers can attempt to construct more height functions using concave-down functions. The goal is to come up with a positive valued function $f(n)$ of all positive integers n such that, for example, $f(n) \cdot f(n+3) < f(n+1) \cdot f(n+2)$ for all such n, or $f(n) + f(n+3) < f(n+1) + f(n+2)$ for all such n.

Example 7. A football of MO brand is covered by some polygonal pieces of leather which are sewed up by three different colours of thread. It features as follows:

(I) Any edge of a polygonal piece of leather is sewed up with an equal-length edge of another polygonal piece of leather by thread of a certain colour.

(II) Each node on the football is vertex to exactly three polygons, and the three threads joint at the node are of different colours.

Show that we can assign to each node on the football a complex number (not equal to 1), such that the product of the numbers assigned to the vertices of any polygonal face is equal to 1 (problem from the 1991 China Mathematical Olympiad).

Solution. Assign each edge a number 1, ω, or ω^2, where $\omega = -\frac{1}{2} + \frac{\sqrt{3}}{2}i$, if the edge is sewed by 2 red, yellow, or blue thread, respectively.

Then assign each node a number ω or ω^2, if the three edges joint at the node have the numbers 1, ω, and ω^2 anticlockwise or not.

For a k-gon with its vertices $A_1, A_2, \ldots, A_k$ anticlockwise, let number ω_i be the value assigned to A_i, z_i be the number assigned to edge $A_{i-1}A_i$, then

$$\omega_i = \frac{z_i}{z_{i+1}}, \quad \text{for } i = 1, 2, \ldots, k, \ (A_0 = A_k, z_{k+1} = z_1, A_{k+1} = A_1)$$

since

$$\omega = \frac{\omega}{1} = \frac{\omega^2}{\omega} = \frac{1}{\omega^2}, \quad \omega^2 = \frac{\omega^2}{1} = \frac{1}{\omega} = \frac{\omega}{\omega^2}.$$

Thus,

$$\omega_1 \omega_2 \cdots \omega_k = \frac{z_1}{z_2} \cdot \frac{z_2}{z_3} \cdots \frac{z_k}{z_1} = 1.$$

Remark. The problem can be stated in an easier way: show that we can assign to each node on the ball a number 1 or 2, such that the sum of the numbers assigned to the vertices of any polygonal face is a multiple of 3.

Example 8. Divide n piles of balls into $n + k$ piles (n, $k > 0$ and each pile has at least 1 ball). Prove that there are at least $k + 1$ balls coming from larger piles.

Solution. Denote M the set of all balls. If a ball is at a pile of a balls, then assign the ball a number $1/a$. In this way, the sum of the values of balls in a pile is 1, and thus, the total sum of the values of all balls is the number of piles. For each ball $A \in M$, let ball A belong to a pile of x_A and y_A balls, before and after the rearrangement, respectively, and let $d(A) = \frac{1}{y_A} - \frac{1}{x_A}$. Since,

$$\sum_{A \in M} d(A) = \sum_{A \in M} \frac{1}{y_A} - \sum_{A \in M} \frac{1}{x_A} = (n+k) - n = k,$$

and $d(A) = \frac{1}{y_A} - \frac{1}{x_A} < \frac{1}{y_A} \le 1$, there are at least $k+1$ balls A_j, such that $d(A_j) = \frac{1}{y_{A_j}} - \frac{1}{x_{A_j}} > 0$, that is, $x_{A_j} > y_{A_j}$.

Exercises

1. There are 11 cups facing up at first. Turn over any even number of them as one *operation*. Show that you cannot have all the 11 cups facing down by any number of *operations*.   ·

2. Let m points in $\triangle ABC$, together with points A, B, and C, form the vertices of several non-overlapping small triangles. Colour points A, B, and C red, yellow, and blue, respectively, and colour each of the m points in $\triangle ABC$ any one of these three colours. Is the number of small triangles whose vertices are in distinct colours odd or even?

3. Divide a square $ABCD$ into n^2 equal small squares (n is a positive integer), colour the opposite vertices A, C red, and B, D blue, and other vertices either red or blue. Prove that the number of small squares with exactly three vertices in a colour is even.

4. There is a grid infinitely towards the right and downward as shown in Fig. 15.6. At first, on the grid, there is only one stone at the upper left square A. Then the rule is: if at some step, there is a stone at P, and there are no stones at its two right neighbouring squares Q and R, or its two downward neighbouring squares S and T, then one can remove the stone at P and put stones at Q and R or at S and T. After infinite steps, no stones are at the first 4 columns of the grid. Is this possible?

Figure 15.6

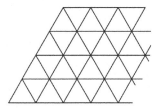

Figure 15.7

5. In a plane, there is an infinite triangular grid consisting of equilateral triangles with unit side lengths as shown in Fig. 15.7. Call the vertices of the triangles the lattice points, call two lattice points with unit distance *adjacent* ones.

A jump game is played by two frogs A and B, 'a *jump*' is called if a frog jumps from the point on which it is lying on to its adjacent point, 'a round *jump*' of A, B is called if first A jumps and then B by the following rules:

Rule (1). A jumps once arbitrarily, then B jumps once in the same direction, or twice in the opposite direction;

Rule (2). When A, B sit on adjacent lattice points, they carry out Rule (1) finishing 'a round *jump*', or 'a *jump*' twice consecutively, keep adjacent with B every time, and B rests on the previous position;

If the original positions of A, B are adjacent lattice points, determine whether for A and B, each one can exactly land on the original position of the other after a finite round of *jumps* (Problem from the 2008 China Team Selection Test).

6. A 3000×3000 square is tiled by dominoes (i.e. 1×2 rectangles) in an arbitrary way. Show that one can colour the dominoes in three colours such that the number of the dominoes of each colour is the same, and each domino d has at most two neighbours of the same colour as d. (Two dominoes are said to be neighbours if a cell of one domino has a common edge with a cell of the other one.) (Problem from the 2006 All-Russian Mathematical Olympiad).

Chapter 16

Calculate in Two Ways

'Calculate in two ways', also called the Fubini principle, is a very important mathematical method. It is so called since one considers (calculates, estimates) the same quantity from two aspects in the process of solving the problem, and obtains an equation which leads to the conclusion. The spirit of this approach is consistent with '*consider the problem in a different perspective*'. In fact, we are familiar with this method, for example, when solving a word problem, we often set a variable, establish two functions, and link them by an equation. That is just to express a quantity in two different ways. In the previous sections (for example, parity, area method, colouring, etc.), the idea of *calculate in two ways* has been touched upon. In this section, we shall focus on some examples and techniques of this method.

Example 1. Prove that

$$\frac{n(n + 1)(n + 2)}{6} = 1 \cdot n + 2 \cdot (n - 1) + 3 \cdot (n - 2) + \cdots + n \cdot 1.$$

Solution. Since the left-hand side of the equality is the combinatorial number S of the different ways of taking any three numbers in numbers $1, 2, \ldots, n + 2$.

That is, $\frac{n(n+1)(n+2)}{6} = S$.

On the other hand, when taking three numbers a, b, c ($1 \leq a < b < c \leq n + 2$) in numbers $1, 2, \ldots, n + 2$, first fix $b \in \{2, 3, \ldots, n + 1\}$, then take a in $1, 2, \ldots, b - 1$, and take c in $b + 1, b + 2, \ldots, n + 2$ arbitrarily. By the multiplication principle, the number of ways of taking the pair (a, c) is $(b-1)(n+2-b)$. Hence,

$$S = \sum_{b=2}^{n+1} (b-1)(n+2-b) = 1 \cdot n + 2 \cdot (n-1) + 3 \cdot (n-2) + \cdots + n \cdot 1.$$

This completes the proof.

Remark. The problem can also be proved by induction. By using the method of Example 1, you can prove a general case as follows:

For $0 < s \le r \le m$, $\sum_{k=0}^{m-r} \binom{s+k-1}{s-1} \binom{m-s-k}{r-s} = \binom{m}{r}$.

(Example 1 is the special case of $m = n+2$, $r = 3$, $s = 2$).

Example 2. Denote $f(n, k)$ the number of factors (no less than k) of positive integers n. Find $\sum_{k=1}^{1000} f(1000 + k, k)$.

Solution. We prove a general conclusion:

$$\sum_{k=1}^{n} f(n + k, k) = 2n, \quad n \in \mathbb{N}^+. \tag{1}$$

For $n, i \in \mathbb{N}^+$, if i is a factor of n, define $g_n(i) = 1$, otherwise $g_n(i) = 0$. Then

$$\sum_{k=1}^{n} f(n + k, k) = \sum_{k=1}^{n} \sum_{i=k}^{n+k} g_{n+k}(i)$$

$$= \sum_{k=1}^{n} \sum_{i=k}^{n} g_{n+k}(i) + \sum_{k=1}^{n} \sum_{i=n+1}^{n+k} g_{n+k}(i) = A + B. \tag{2}$$

Obviously, in i consecutive integers $n+1, n+2, \ldots, n+i$, the number i is a factor of exactly one of these numbers. Exchange the order of summation, and we have

$$A = \sum_{i=1}^{n} \sum_{k=1}^{i} g_{n+k}(i) = \sum_{i=1}^{n} 1 = n. \tag{3}$$

And if $1 \le k \le n$, in numbers $n+1, n+2, \ldots, n+k$ only $n+k$ is a factor of $n+k$. So

$$B = \sum_{k=1}^{n} \sum_{i=n+1}^{n+k} g_{n+k}(i) = \sum_{k=1}^{n} 1 = n. \tag{4}$$

Substitute (3), (4) into (2), and then (1) follows.

Example 3. Show that you cannot find n (n is even) entries with values $1, 2, \ldots, n$ in the $n \times n$ matrix $\begin{pmatrix} 1 & 2 & 3 & \cdots & n \\ 2 & 3 & 4 & \cdots & 1 \\ 3 & 4 & 5 & \cdots & 2 \\ \cdots & \cdots & \cdots & & \cdots \\ n & 1 & 2 & \cdots & n-1 \end{pmatrix}$, such that any two of them are neither at the same row nor at the same column.

Solution. Suppose, on the contrary, there were n entries in the matrix, such that each row and each column of the matrix had exactly one of the entries. Let the entry with value k, ($k = 1, \ldots, n$) be on the i_k row and the j_k column. Since n is even,

$$\sum_{k=1}^{n} i_k = \sum_{k=1}^{n} j_k = \frac{n(n+1)}{2} \equiv \frac{n}{2} \pmod{n}.$$

On the other hand, by the configuration of the matrix, we have

$$i_k + j_k - 1 \equiv k \pmod{n}.$$

Hence,

$$\sum_{k=1}^{n} i_k + \sum_{k=1}^{n} j_k - \sum_{k=1}^{n} 1 \equiv \sum_{k=1}^{n} k \pmod{n}.$$

That is, $\frac{n}{2} + \frac{n}{2} - n \equiv \frac{n}{2} \pmod{n}$, a contradiction.

Example 4. For given $n(n \geq 3)$ real numbers $a_1, a_2, \ldots, a_n$, denote the minimum of $|a_i - a_j|$ ($1 \leq i < j \leq n$) by m. Find the maximum of m subject to the condition $a_1^2 + a_2^2 + \cdots + a_n^2 = 1$.

Solution. Without loss of generality, assume that $a_1 \leq a_2 \leq \cdots \leq a_n$. We estimate $S = \sum_{1 \leq i < j \leq n} (a_j - a_i)^2$ in two ways.

On the one hand,

$$S = (n-1)\sum_{i=1}^{n} a_i^2 - 2 \sum_{1 \leq i < j \leq n} a_i a_j = n \cdot \sum_{i=1}^{n} a_i^2 - \left(\sum_{i=1}^{n} a_i\right)^2 \leq n.$$

On the other hand, since $a_{i+1} - a_i \geq m$ for $i = 1, \ldots, n-1$, we have $a_j - a_i \geq (j-i)m$ for $1 \leq i < j \leq n$. Thus,

$$S \geq m^2 \sum_{1 \leq i < j \leq n} (j-i)^2 = m^2 \left(n \cdot \sum_{i=1}^{n} i^2 - \left(\sum_{i=1}^{n} i\right)^2 \right) = \frac{m^2}{12} n^2 (n^2 - 1).$$

Hence, $n \geq \frac{m^2}{12}n^2(n^2-1)$, therefore,

$$m \leq \sqrt{\frac{12}{n(n^2-1)}}.$$

If $\sum_{i=1}^{n} a_i = 0$ and if $a_1, a_2, \ldots, a_n$ is an arithmetic sequence, equality holds, thus, the maximum of m is $\sqrt{\frac{12}{n(n^2-1)}}$.

Example 5. There are $n(n \geq 2)$ finite sets $A_1, A_2, \ldots, A_n$ with $|A_i \cap A_j| \leq 1$, for $1 \leq i < j \leq n$, and for any $a \in M = \bigcup_{i=1}^{n} A_i$, a belongs to at least two sets of $A_1, A_2, \ldots, A_n$. Find the maximum of $|M|$. ($|X|$ is the number of elements in the set X.)

Solution. Suppose that $M = \{a_1, a_2, \ldots, a_m\}$, where a_k $(k = 1, 2, \ldots, m)$ is in t_k sets of $A_1, A_2, \ldots, A_n$, $t_k \geq 2$.

First, consider the number S of the triples (a_k, A_i, A_j) with $a_k \in A_i \cap A_j$, for $1 \leq k \leq m, 1 \leq i < j \leq n$.

On the one hand, summing up first on k, then on i, j, we have

$$S = \sum_{1 \leq i < j \leq n} |A_i \cap A_j| \leq \sum_{1 \leq i < j \leq n} 1 = \frac{n(n-1)}{2}.$$

On the other hand, summing up first on i, j, then on k, we have

$$S = \sum_{k=1}^{m} t_k(t_k - 1) \bigg/ 2 \geq \sum_{k=1}^{m} 1 = m. \qquad (5)$$

Therefore, $|M| = m \leq \frac{n(n-1)}{2}$.

Now we construct an example such that $|M| = \frac{n(n-1)}{2}$.

Denote all binary subsets of $\{1, 2, \ldots, n\}$ by $B_1, B_2, \ldots, B_N$, $N = \frac{n(n-1)}{2}$. Let $a_k \in A_i$ if and only if $i \in B_k$, $k = 1, 2, \ldots, N$. Since $|B_k| = 2$, each a_k is in two sets of $A_1, A_2, \ldots, A_n$. Furthermore, if $a_k \in A_i \cap A_j$, then $i, j \in B_k$, such k is unique. Hence, $|A_i \cap A_j| \leq 1$. So, the maximum of $|M|$ is $\frac{n(n-1)}{2}$.

Remark. The method can also be used to find the maximum of $\sum_{i=1}^{n} |A_i|$. We see that

$$S = \sum_{k=1}^{m} \frac{t_k(t_k - 1)}{2} \geq \frac{1}{2} \sum_{k=1}^{m} t_k = \frac{1}{2} \sum_{i=1}^{n} |A_i|.$$

Where the last equality is obtained by considering the number of pairs (a_k, A_i) with $a_k \in A_i$, the left-hand side of the equality is obtained by summing up first on i, then on k; the right-hand side of the equality is obtained by summing up first on k then on i. Hence, we have, $\sum_{i=1}^{n} |A_i| \leq n(n-1)$. Equality holds for the above example. So, the maximum of $\sum_{i=1}^{n} |A_i|$ is $n(n-1)$.

Example 6. Let A be a set with $|A| = 225$, meaning that A has 225 elements. Suppose further that there are 11 subsets $A_1, A_2, \ldots, A_{11}$ of A with $|A_i| = 45$ for $1 \leq i \leq 11$ and $|A_i \cap A_j| = 9$ for $1 \leq i < j \leq 11$. Prove that $|A_1 \cup A_2 \cup \cdots \cup A_{11}| \geq 165$, and give an example for which equality holds (Problem from the 2011 USA Mathematical Olympiad).

Solution. Let $X = A_1 \cup A_2 \cup \cdots \cup A_{11}$, $f_i(x) = \begin{cases} 1, & \text{if } x \in A_i \\ 0, & \text{if } x \notin A_i \end{cases}$, $1 \leq i \leq 11$. Obviously, $f_i(x) = f_i^2(x)$. Let $d(x) = \sum_{i=1}^{11} f_i(x)$, which means x is in $d(x)$ sets of $A_1, A_2, \ldots, A_{11}$.

On the one hand,

$$\sum_{x \in X} d^2(x) = \sum_{x \in X} \sum_{i=1}^{11} f_i^2(x) + 2 \sum_{x \in X} \sum_{1 \leq i < j \leq 11} f_i(x) f_j(x)$$

$$= \sum_{i=1}^{11} \sum_{x \in X} f_i(x) + 2 \sum_{1 \leq i < j \leq 11} \sum_{x \in X} f_i(x) f_j(x)$$

$$= \sum_{i=1}^{11} |A_i| + 2 \sum_{1 \leq i < j \leq 11} |A_i \cap A_j|$$

$$= 11 \times 45 + 2 \times \binom{11}{2} \times 9$$

$$= 1485.$$

On the other hand,

$$\sum_{x \in X} d^2(x) \geq \frac{1}{|X|} \left(\sum_{x \in X} d(x) \right)^2 = \frac{1}{|X|} \left(\sum_{x \in X} \sum_{i=1}^{11} f_i(x) \right)^2$$

$$= \frac{1}{|X|} \left(\sum_{i=1}^{11} |A_i| \right)^2 = \frac{(11 \times 45)^2}{|X|}.$$

Hence, $1485 \geq \frac{(11 \times 45)^2}{|X|}$, that is, $|X| \geq 165$.

An example such that equality holds is as follows.

Let the elements of A be all triples of $\{1, 2, \ldots, 11\}$ and any other 60 elements, totally $\binom{11}{3} + 60 = 225$ elements.

For $1 \leq i \leq 11$, let the elements of A_i be all triples of $\{1, 2, \ldots, 11\}$ containing i, then $|A_i| = \binom{10}{2} = 45$. For any $1 \leq i < j \leq 11$, $|A_i \cap A_j| = \binom{9}{1} = 9$. And,

$$|A_1 \cup A_2 \cup \cdots \cup A_{11}| = \binom{11}{3} = 165.$$

Example 7. Let S be a finite set of points in the plane such that no three of them are on a line. For each convex polygon P whose vertices are in S, let $a(P)$ be the number of vertices of P, and let $b(P)$ be the number of points of S which are outside P. A line segment, a point, and the empty set are considered as convex polygons of 2, 1, and 0 vertices, respectively. Prove that for every real number x,

$$\sum_P x^{a(P)} (1 - x)^{b(P)} = 1,$$

where the sum is taken over all convex polygons P with vertices in S (Problem from the 2006 International Mathematical Olympiad Shortlist).

Solution. Let $|S| = n$. For each convex polygon P with vertices in S, denote the number of points in S which are inside P by $c(P)$, then $a(P) + b(P) + c(P) = n$.

Denote $1 - x = y$, then,

$$\sum_P x^{a(P)} (1 - x)^{b(P)} = \sum_P x^{a(P)} y^{b(P)} = \sum_P x^{a(P)} y^{b(P)} (x + y)^{c(P)}$$

$$= \sum_P \sum_{i=0}^{c(P)} \binom{c(p)}{i} x^{a(P)+i} y^{b(P)+c(P)-i}. \tag{6}$$

This is an nth order homogeneous polynomial of x and y.

Fix $r (0 \leq r \leq n)$, and consider the coefficient of $x^r y^{n-r}$ in (6).

Select a convex polygon P with $a(P) \leq r$, and take $i = r - a(P)$ points of $c(P)$ points inside P.

On the one hand, there are $\sum_P \binom{c(P)}{r - a(P)}$ ways of selecting, which is exactly the coefficient of $x^r y^{n-r}$ in (6).

On the other hand, such number of ways corresponds to the number of ways to select r vertices in S, that is $\binom{n}{r}$. The correspondence is one to one, since each subset T of S can be uniquely divided into two disjoint subsets, one is the convex hull of T, and the other is the points of T within the hull.

Summing up over P, the coefficient of $x^r y^{n-r}$ in (6) is $\sum_P \binom{c(P)}{r - a(P)} = \binom{n}{r}$, thus, $\sum_P x^{a(P)}(1 - x)^{b(P)} = \sum_{r=0}^{n} \binom{n}{r} x^r y^{n-r} = (x + y)^n = 1$.

Remark. If we only consider the convex polygons in the usual sense, that is, the number of vertices $a(P) \geq 3$, then the result should be

$$\sum_P x^{a(P)}(1 - x)^{b(P)} = 1 - \sum_{k=0}^{2} \binom{n}{k} x^k (1 - x)^{n-k}.$$

Exercises

1. Prove that there does not exist a sequence of 17 terms such that the sum of any each seven consecutive terms is positive, and the sum of any 11 consecutive terms is negative.

2. Let $d(n)$ be the number of positive factors of the positive integer n. Prove that:

$$\sum_{k=1}^{n} d(k) = \sum_{k=1}^{n} \left\lfloor \frac{n}{k} \right\rfloor.$$

3. In a school there are b teachers and c students. Suppose that

   (a) each teacher teaches exactly k students, and
   (b) for any two (distinct) students, exactly h teachers teach both of them.

   Prove that $b/h = c(c-1)/(k(k-1))$ (Problem from the 2004 Hong Kong Mathematical Olympiad) (see also Example 5 of Chapter 20).

4. Prove that one can always find 2 people in a group of 12 people, such that there are at least 5 persons among the other 10 persons who each know or do not know these 2 people.

5. In a rectangular array of non-negative real numbers with m rows and n columns, each row and each column contains at least one positive element. Moreover, if a row and a column intersect in a positive element, then the sums of their elements are the same. Prove that $m = n$ (Problem from the 2006 Canadian Mathematical Olympiad).

6. Let $x_1, x_2, \ldots, x_n$ and $y_1, y_2, \ldots, y_n$ be real numbers. Let $A = (a_{ij})_{1 \leq i,j \leq n}$ be the matrix with entries

$$a_{ij} = \begin{cases} 1, & \text{if } x_i + y_j \geq 0, \\ 0, & \text{if } x_i + y_j < 0. \end{cases}$$

Suppose that B is an $n \times n$ matrix with entries 0, 1 such that the sum of the elements in each row and each column of B is equal to the corresponding sum for the matrix A. Prove that $A = B$ (Problem from the 2003 International Mathematical Olympiad Shortlist).

Chapter 17

Stepwise Adjustment

The so-called stepwise adjustment method involves fixing some variable of a problem temporarily at first, then studying the influence of other variables on the solution of the problem. When obtaining some partial results, try to guess the result of the whole problem.

The spirit of stepwise adjustment method is reflected in the following aspects:

1. Adjust the original state to a better state in some sense, so as to gradually achieve the optimal state;
2. Adjust a large number of general cases to a few special, regular cases, so as to make the problem easier;
3. Follow some intrinsic rules in the operation process (such as invariance, parity, congruence, continuity, monotonicity, etc.) and make gradual adjustments, so as to draw a certain conclusion or prove a proposition;
4. Once some restrictions are satisfied, make further adjustments such that more restrictions are satisfied.

Let us look at some concrete examples.

Example 1. Let $x_1, x_2, \ldots, x_{40}$ be positive integers, and $\sum_{i=1}^{40} x_i = 58$. Find the sum of the maximum A and the minimum B of $\sum_{i=1}^{40} x_i^2$.

Solution. Since the number of ways of writing 58 as the sum of 40 positive integers is finite, A and B exist.

Without loss of generality, let $x_1 \leq x_2 \leq \cdots \leq x_{40}$. If $x_1 > 1$, since $x_1 + x_{40} = (x_1 - 1) + (x_{40} + 1)$ and $(x_1 - 1)^2 + (x_{40} + 1)^2 = x_1^2 + x_{40}^2 + 2(x_{40} - x_1) + 2 > x_1^2 + x_{40}^2$, we can adjust x_1 step by step to 1, and $\sum_{i=1}^{40} x_i^2$ will increase. In the same way, we can adjust $x_2, x_3, \ldots, x_{39}$ successively to 1, then $\sum_{i=1}^{40} x_i^2$ takes its maximum A. That is, when $x_1, x_2, \ldots, x_{39} = 1$, $x_{40} = 19$, the maximum A of $\sum_{i=1}^{40} x_i^2$ is 400.

To achieve the minimum B of $\sum_{i=1}^{40} x_i^2$, $x_1, x_2, \ldots, x_{40}$ should be as close as possible, since if there are two adjacent numbers x_i, x_{i+1}, $(1 \le i \le 39)$, with $x_{i+1} - x_i \ge 2$, then,

$$(x_i + 1)^2 + (x_{i+1} - 1)^2 = x_i^2 + x_{i+1}^2 - 2(x_{i+1} - x_i - 1) < x_i^2 + x_{i+1}^2.$$

Since the average of $x_1, x_2, \ldots, x_{40}$ is $58/40 = 1.45$, we see that $x_i = 1$ or 2. Let $x_1 = x_2 = \cdots = x_{22} = 1$, $x_{23} = x_{24} = \cdots = x_{40} = 2$, then $\sum_{i=1}^{40} x_i^2$ takes its minimum $B = 94$. Therefore, $A + B = 494$.

Example 2. Let 2006 be expressed as the sum of five positive integers x_1, x_2, x_3, x_4, x_5, and $S = \sum_{1 \le i < j \le 5} x_i x_j$.

(a) What value of $(x_1, x_2, x_3, x_4, x_5)$ maximizes S?

(b) Find, with proof, the value of $(x_1, x_2, x_3, x_4, x_5)$ which minimizes of S if $1 \le i, j \le 5$, $|x_i - x_j| \le 2$ for any $1 \le i, j \le 5$ (Problem from the 2006 China High School Mathematical League).

Solution. (a) Since S is bounded and its domain is finite, S attains its maximum and minimum. Suppose that $x_1 \ge x_2 \ge x_3 \ge x_4 \ge x_5 \ge 1$. If $S = \sum_{1 \le i < j \le 5} x_i x_j$ attains its maximum, then we must have

$$0 \le x_1 - x_5 \le 1. \tag{1}$$

In fact, if (1) does not hold, then, $x_1 - x_5 \ge 2$. Let $x_1' = x_1 - 1$, $x_5' = x_5 + 1$, $x_i' = x_i (i = 2, 3, 4)$, we have $x_1' + x_5' = x_1 + x_5$, $x_1' \cdot x_5' = x_1 x_5 + x_1 - x_5 - 1 > x_1 x_5$.

Denote $S = \sum_{1 \le i < j \le 5} x_i x_j$, $S' = \sum_{1 \le i < j \le 5} x_i' x_j'$, then, $S' - S = x_1' x_5' - x_1 x_5 > 0$. A contradiction.

So, we can always adjust as above such that $0 \le x_1 - x_5 \le 1$ to obtain the maximum of S. Thus, when $x_1 = 402, x_2 = x_3 = x_4 = x_5 = 401$, S attains its maximum.

(b) When $x_1 + x_2 + x_3 + x_4 + x_5 = 2006$ and $0 \le x_1 - x_5 \le 2$, there are only three cases:

   (I) 402, 402, 402, 400, 400;
   (II) 402, 402, 401, 401, 400;
   (III) 402, 401, 401, 401, 401;

In the proof of (a), we see that case (II) is obtained by adjusting x_3, x_4 of case (I). Case (III) is obtained by adjusting x_2, x_5 of case (II). So, S attains its minimum in case (I).

Example 3. Let A, B, and C be the angles of $\triangle ABC$. Find the maximum of $\sin \frac{A}{2} \sin \frac{B}{2} \sin \frac{C}{2}$.

Solution. First fix A temporarily, let $u = \sin \frac{A}{2} \sin \frac{B}{2} \sin \frac{C}{2}$. Then

$$u = \sin \frac{A}{2} \times \frac{1}{2} \left(\cos \frac{B-C}{2} - \cos \frac{B+C}{2} \right)$$

$$= \frac{1}{2} \sin \frac{A}{2} \left(\cos \frac{B-C}{2} - \sin \frac{A}{2} \right).$$

Obviously, when $B = C$, $\cos \frac{B-C}{2}$ attains its maximum 1. Hence,

$$u \le \frac{1}{2} \sin \frac{A}{2} \left(1 - \sin \frac{A}{2} \right) = \frac{1}{8} - \frac{1}{2} \left(\sin \frac{A}{2} - \frac{1}{2} \right)^2 \le \frac{1}{8}.$$

When $A = B = C = 60°$, u attains its maximum, $1/8$.

Example 4. Find a point P in or at the sides of $\triangle ABC$ such that the sum of the distances of P to the three vertices is maximal.

Solution. Denote $f(P') = P'A + P'B + P'C$.

First, we show that the point P cannot be an inner point of $\triangle ABC$.

In fact, if P were an inner point of $\triangle ABC$, make an ellipse passing through point P with focuses on B and C. Let the ellipse intersect AB and AC at points P_1 and P_2, respectively. Extend AP to a point P' on $P_1 P_2$ (see Fig. 17.1). Then $P'A < \max\{P_1A, P_2A\}$.

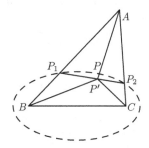

Figure 17.1

Without loss of generality, suppose that $P'A < P_1A$, there would be

$$f(P_1) = P_1A + P_1B + P_1C > PA + (P_1B + P_1C)$$
$$= PA + (PB + PC) = f(P).$$

A contradiction.

Second, we show that the point P must be one of the vertices.

If P were on BC but not the endpoint B or C, then $PA < \max\{BA, CA\}$, say $PA < BA$, there would be $f(B) = BA + BC > PA + (PB + PC) = f(P)$.

A contradiction.

Finally, it is easy to see that P coincides with the vertex of the smallest angle of $\triangle ABC$.

Example 5. Find the minimal real number λ such that

$$5(abc + abd + acd + bcd) \le \lambda abcd + 12$$

holds for any positive real numbers a, b, c and d with $a + b + c + d = 4$.

Solution. Taking $a = b = c = d = 1$, we have $\lambda \ge 8$. We shall prove that the minimal of λ is 8.

That is, we shall prove that

$$f(a, b, c, d) = 5(abc + abd + acd + bcd) - 8abcd$$
$$= ab(5c + 5d - 8cd) + 5cd(a + b) \le 12.$$

If $5c + 5d \le 8cd$, then

$$f(a, b, c, d) \le 5cd(a + b) \le 5\left(\frac{(a + b) + c + d}{3}\right)^3 = 5\left(\frac{4}{3}\right)^3 < 12.$$

Otherwise,

$$f(a, b, c, d) = ab(5c + 5d - 8cd) + 5cd(a + b)$$
$$\le \left(\frac{a + b}{2}\right)^2 (5c + 5d - 8cd) + 5cd(a + b)$$
$$= f\left(\frac{a + b}{2}, \frac{a + b}{2}, c, d\right).$$

Repeatedly using the symmetry of the variables of f and the above inequality, we have

$$f(a, b, c, d) \leq f\left(\frac{a+b}{2}, \frac{a+b}{2}, c, d\right)$$

$$\leq f\left(\frac{a+b}{2}, \frac{a+b}{2}, \frac{c+d}{2}, \frac{c+d}{2}\right)$$

$$\leq f\left(\frac{a+b}{2}, 1, 1, \frac{c+d}{2}\right) \leq f(1, 1, 1, 1) = 12.$$

Example 6. Let S be a set of $6n$ points in a line. Choose randomly $4n$ of these points and paint them blue; the other $2n$ points are painted green. Prove that there exists a line segment that contains exactly $3n$ points of S, $2n$ of them blue and n of them green (Problem from the 2008 Brazil National Olympiad).

Solution. Denote these $6n$ points from left to right by $A_1, A_2, \ldots, A_{6n}$.

Define $f(i)$ the number of blue points in $A_i, A_{i+1}, \ldots, A_{i+3n-1}$, $i = 1, 2, \ldots, 3n + 1$. Suppose that $f(1) \geq 2n$, otherwise, denote from right to left. It suffices to show there is a j such that $f(j) = 2n$.

Since $f(1) + f(3n + 1) = 4n$,

$$f(3n + 1) \leq 2n \leq f(1). \tag{1}$$

We see that, for $i = 1, 2, \ldots, 3n$,

$$f(i + 1) = \begin{cases} f(i), \text{ if } A_i \text{ and } A_{i+3n} \text{ are in the same colour;} \\ f(i) - 1, \text{ if } A_i \text{ is blue and } A_{i+3n} \text{ is green;} \\ f(i) + 1, \text{ if } A_i \text{ is green and } A_{i+3n} \text{ is blue.} \end{cases} \tag{2}$$

By (1) and (2), we see that there exists at least one $j \in \{1, 2, \ldots, 3n + 1\}$, such that $f(j) = 2n$, which completes the proof.

Example 7. Let $1^2, 2^2, \ldots, n^2$ be randomly placed on a circle. Define an *action* as adding $a \in \mathbb{N}^+$ to two adjacent numbers while keeping all the other numbers unchanged. Find all integers $n \geq 3$, such that the n numbers can always be the same after no more than n *actions*.

Solution. Let the numbers be $a_1, \ldots, a_n$ clockwise, and use the convention of $a_{i+n} = a_i$.

Suppose that after several *actions*, all numbers are equal to S. We first show that n cannot be even.

If n were even, denote the sum of numbers $a_1, a_3, \ldots, a_{n-1}$ by P, and the sum of numbers $a_2, a_4, \ldots, a_n$ by Q. Under each action, the increments of P and Q are equal. Thus $P - Q$ is unchanged. Finally, $P - Q = 0$. Thus, at initial time, $P - Q = a_1 - a_2 + a_3 - \cdots + a_{n-1} - a_n = 0$. Take $(a_1, a_2, \ldots, a_n) = (1^2, 2^2, \ldots, n^2)$, then $a_1 - a_2 + a_3 - \cdots + a_{n-1} - a_n < 0$, a contradiction.

Now we shall prove a stronger result: if $n(\geq 3)$ is odd, any array of positive integers $a_1, a_2, \ldots, a_n$ are equal to a number S after no more than n *actions*.

If we take *actions* to a_i and a_{i+1}, a_{i+2} and $a_{i+3}, \ldots, a_{i+n-1}$ and $a_{i+n}(= a_i)$ with $a = 1$, then a_i increases by 2 and others increase by 1. We call such $\frac{n+1}{2}$ actions a *group of actions* to a_i.

Let $a_j = \max\{a_1, a_2, \ldots, a_n\}$. We make $a_j - a_i$ *groups of actions* to $a_i, i = 1, 2, \ldots, n$, all together $\sum_{i=1}^{n}(a_j - a_i)$ *groups of actions*. Then a_i changes to the number

$$a_i + 2(a_j - a_i) + \sum_{\substack{k=1 \\ k \neq i}}^{n}(a_j - a_k)$$

$$= a_i + (a_j - a_i) + \sum_{k=1}^{n}(a_j - a_k)$$

$$= a_j + \sum_{k=1}^{n}(a_j - a_k) =: S,$$

$$i = 1, 2, \ldots, n.$$

Regard these $\sum_{i=1}^{n}(a_j - a_i)$ *groups of actions* as $\frac{n+1}{2}\sum_{k=1}^{n}(a_j - a_k)$ *actions* with $a = 1$. Then combine all $q_i = \frac{n+1}{2}(a_j - a_i)$ *actions* to one *action* to a_i and a_{i+1} with $a = q_i$. Thus, all together no more than n such *actions* will change all numbers to S.

Remark. Here is another solution. Suppose for $i = 1, 2, \ldots, n, n \geq 3$ is odd, b_i is added to a_i, a_{i+1}, and all the numbers become positive integers M after n *actions*. We have $b_{i-1} + a_i + b_i = M$ for $i = 1, 2, \ldots, n$. To solve for b_i, add the two equations involving b_i and alternately add and subtract the other equations. We have $2b_i = M - a_i - a_{i+1} + (a_{i+2} - a_{i+3} + \cdots - \cdots + a_{i-1})$. For suitably chosen large M, all b_is are positive integers and the proof is complete.

Example 8. Let a, b, c be non-zero rational numbers such that equation $ax^2 + by^2 + cz^2 = 0$ possesses a non-zero rational number solution (x_0, y_0, z_0). Show that for any $N > 0$, there exists a rational number solution (x_1, y_1, z_1) such that $x_1^2 + y_1^2 + z_1^2 > N$.

Solution. We have $ax_0^2 + by_0^2 + cz_0^2 = 0$, and say $x_0 \neq 0$. Let $x = k + tx_0$, $y = ty_0$, $z = tz_0$, where $k > 0$ and t are rational numbers to be determined. Then

$$ax^2 + by^2 + cz^2 = a(k + tx_0)^2 + bt^2 y_0^2 + ct^2 z_0^2$$
$$= ak^2 + 2akx_0 t + t^2(ax_0^2 + by_0^2 + cz_0^2)$$
$$= ak^2 + 2akx_0 t.$$

Take t such that $ak^2 + 2akx_0 t = 1$, that is, $t = \frac{1 - ak^2}{2akx_0}$, then

$$\begin{cases} x_1 = k + tx_0 = \frac{k}{2} + \frac{1}{2ak}, \\ y_1 = ty_0 = \frac{1 - ak^2}{2ak} \cdot \frac{y_0}{x_0}, \\ z_1 = tz_0 = \frac{1 - ak^2}{2ak} \cdot \frac{z_0}{x_0} \end{cases}$$

is a rational number solution of $ax^2 + by^2 + cz^2 = 1$ (where k is still a parameter to be determined).

Take a rational number $k > \max\left\{2\sqrt{N} + 1, \frac{1}{|a|}\right\}$, then

$$\frac{k}{2} > \sqrt{N} + \frac{1}{2}, \left|\frac{1}{2ak}\right| < \frac{1}{2|a|} \cdot |a| = \frac{1}{2}.$$

Thus, $x_1 = \frac{k}{2} + \frac{1}{2ak} > \sqrt{N} + \frac{1}{2} - \frac{1}{2} = \sqrt{N}$, and $x_1^2 + y_1^2 + z_1^2 > N$.

Example 9. Let $A_1, A_2, \ldots, A_n$ be n non-empty subsets of set X, and for any $i, j \in \{1, 2, \ldots, n\}$, $A_i \cap A_j$ is not a singleton. Show that: we can divide the elements of set X into two classes, such that the elements of each subset A_i $(i = 1, 2, \ldots, n)$ are not in the same class.

Solution. Taking $j = i$, each subset A_i has at least two elements. If the elements of A_i are all in one class, we call A_i a *single class set*, otherwise we call A_i a *two-class set*. Denote t the number of *two-class sets* in A_i $(i = 1, 2, \ldots, n)$.

Since A_1 has at least two elements, we can divide X into two classes, the first class and the second class, such that A_1 is a *two-class set*. So $t \geq 1$.

We show that for any classification of the elements of X such that $t < n$, we can always adjust the classification to increase t.

Let $x_1, x_2, \ldots, x_r$ be in the first class, and $x_{r+1}, x_{r+2}, \ldots, x_m$ be in the second class. And let $A_1, A_2, \ldots, A_t$ be *two-class sets*, and $A_{t+1}, A_{t+2}, \ldots, A_n$ be *single class sets*. Let $A_{t+1} = \{x_1, x_2, \ldots, x_s\}$, $2 \leq s \leq r$.

Let x_1 be in the second class, with other elements unchanged. So A_{t+1} becomes a *two-class set*.

Next, we show that $A_i (1 \leq i \leq t)$ is still a *two-class set*.

Since if $x_1 \notin A_i$, A_i is a *two-class set*; If $x_1 \in A_i$, since $A_i \cap A_{t+1}$ is not a singleton, there is an element $x_j (2 \leq j \leq s)$ of the first class in A_i. That is, A_i is still a *two-class set*.

Exercises

1. Let non-negative real numbers $x_1, x_2, \ldots, x_n$ satisfy $x_1 + x_2 + \cdots + x_n \leq \frac{1}{2}$. Find the minimum of $(1 - x_1)(1 - x_2) \cdots (1 - x_n)$.
2. Let positive integers $x_1, x_2, \ldots, x_{10}$ satisfy

$$x_1 + x_2 + \cdots + x_{10} = 49.$$

   Find the maximum and minimum of $x_1^2 + x_2^2 + \cdots + x_{10}^2$.
3. Partition 2006 into several distinct positive integers such that their product is the largest. What is the largest product?
4. Denote $\pi(n)$ the number of primes not greater than n. Prove that for any positive integer n and non-negative integer $k \leq \pi(n)$, there exist n consecutive positive integers in which there are exactly k primes.
5. For a set of conditions

$$\begin{cases} x_1 + x_2 + \cdots + x_n = m, \\ x_1^2 + x_2^2 + \cdots + x_n^2 - \frac{m^2}{n} < 2. \end{cases} \quad (*)$$

   (a) Find all positive integers n, such that $(*)$ has integer solution $(x_1, x_2, \ldots, x_n)$ for any positive integer m;

   (b) Find all positive integer pairs (m, n), such that $(*)$ has integer solution $(x_1, x_2, \ldots, x_n)$.

6. Given 1989 points in the space, any three of them are not collinear. We divide these points into 30 groups such that the numbers of points in these groups are different from each other. Consider those triangles whose vertices are points belonging to three different groups among

the 30. Determine the numbers of points of each group such that the number of such triangles attains a maximum (Problem from the 1989 China Mathematical Olympiad).

7. The positive reals a, b, c, d satisfy $abcd = 1$. Prove that

$$\frac{1}{a} + \frac{1}{b} + \frac{1}{c} + \frac{1}{d} + \frac{9}{a+b+c+d} \geq \frac{25}{4}.$$

(Problem from the 2011 China Girls' mathematical Olympiad.)

Chapter 18

Methods of Construction

In the process of solving problems, due to some needs, we can either construct the relations in the conditions, or assume these relations to be realized in a model, or construct a new form of the conditions through proper logical combination, so as to arrive at the final conclusion. In this process, creative thinking is characterized by 'construction', which we may call constructive thinking, and the method of solving problems by using constructive thinking is called the method of construction.

To effectively use the constructive method, we need to have comprehensive knowledge and keen intuition, make associations from multiple angles and ways, and integrate the knowledge of algebra, trigonometry, geometry, and number theory into problem solving.

Example 1. Prove that there are infinitely many solutions of positive integers to the Diophantine equation $x^2 + y^2 = z^6$.

Solution. We know that there are infinitely many solutions of positive integers (a, b, c) to $a^2 + b^2 = c^2$. Multiplying both sides by c^4, we have $(ac^2)^2 + (bc^2)^2 = c^6$, and hence, $x = ac^2, y = bc^2, z = c$ is a solution of the desired equation.

Remark. We can prove Example 1 even by just one solution of $a^2 + b^2 = c^2$, say $a = 3$, $b = 4$, and $c = 5$.

Multiply both sides of $a^2 + b^2 = c^2$ by $c^{6k+4} (k \in \mathbb{N})$, and we have

$$(ac^{3k+2})^2 + (bc^{3k+2})^2 = (c^{k+1})^6.$$

To prove this kind of problem as Example 1, we need to construct an identity and use it to generate infinitely many identities of the same type. This is often not so easy to find.

In general, solving these problems requires some understanding and intuition of the various algebraic structures. Although the ideas may vary from person to person, the construction process also has twists and turns now and then, but one should always preserve and follow the original characteristics of the problem. And one should constantly try to come up with good ideas until he/she succeeds. A set of questions are in Exercise 1 to enhance the reader's further understanding.

The construction method is not limited to creating examples; sometimes the construction of equations, functions, polynomials, geometric figures, some algorithm, etc., will become the key step to achieving the transformation of the problems. Here are some examples.

Example 2. Let three real numbers a, b, and c satisfy $(a+c)(a+b+c) < 0$. Prove that

$$(b - c)^2 > 4a(a + b + c).$$

Analysis. We associate $(b - c)^2 > 4a(a+b+c)$ with the discriminant of a quadratic trinomial $y = ax^2 + (b - c)x + (a + b + c)$. And the problem can be solved by studying its graph and roots.

Solution. If $a = 0$, then $c(b + c) < 0$, thus $b \neq c$ (otherwise $2b^2 < 0$). Hence, $(b - c)^2 > 0$. The proposition is true.

If $a \neq 0$, construct a function $y = ax^2 + (b - c)x + (a + b + c)$. We see that $y(-1)y(0) = 2(a+c)(a+b+c) < 0$, hence, the quadratic function has two unequal real roots. Thus, $\Delta > 0$, that is,

$$(b - c)^2 > 4a(a + b + c).$$

Example 3. Let a, b, and c be real numbers with absolute value less than 1. Prove that $ab + bc + ca + 1 > 0$.

Solution. Construct a linear function of a: $f(a) = (b + c)a + bc + 1$, $a \in (-1, 1)$. Its graph is a segment on the plane. Since $f(-1) = -(b+c)+bc+1 = (b - 1)(c - 1) > 0$, and $f(1) = (b+c) + bc + 1 = (b+1)(c+1) > 0$, the segment lies above the x-axis, that is, for all a in $[-1, 1]$, we have $f(a) = (b + c)a + bc + 1 > 0$.

Example 4. Let $a \geq c$, $b \geq c$, $c > 0$. Show that

$$\sqrt{c(a - c)} + \sqrt{c(b - c)} \leq \sqrt{ab}.$$

Solution. If $a = c$ or $b = c$, the inequality holds obviously.

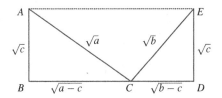

Figure 18.1

If $a > c$ and $b > c$, construct a rectangle as Fig. 18.1, where $AB = DE = \sqrt{c}$, $BC = \sqrt{a-c}$, $CD = \sqrt{b-c}$, $\angle ABC = \angle CDE = \frac{\pi}{2}$. Let $\angle ACE = \alpha$. Then,

$$S_{ABDE} = \sqrt{c}(\sqrt{a-c} + \sqrt{b-c}) = S_{\triangle ABC} + S_{\triangle CDE} + S_{\triangle ACE}$$

$$= \frac{1}{2}\sqrt{c}\sqrt{a-c} + \frac{1}{2}\sqrt{c}\sqrt{b-c} + \frac{1}{2}\sqrt{a}\sqrt{b}\sin\alpha$$

$$\leq \frac{1}{2}\sqrt{c}(\sqrt{a-c} + \sqrt{b-c}) + \frac{1}{2}\sqrt{ab}.$$

Hence, $\sqrt{c(a-c)} + \sqrt{c(b-c)} \leq \sqrt{ab}$. This completes the proof.

Example 5. Let real numbers x, y, z satisfy $0 < x < y < z < \frac{\pi}{2}$. Prove that

$$\frac{\pi}{2} + 2\sin x \cos y + 2\sin y \cos z > \sin 2x + \sin 2y + \sin 2z.$$

Solution. Draw a 1/4 unit circle with centre at the origin in the first quadrant of the plane. Take points A_1, A_2, A_3 on the circle such that $\angle A_1 Ox = x$, $\angle A_2 Ox = y$, $\angle A_3 Ox = z$ as in Fig. 18.2. Since the sum of the areas of three rectangles $S_1 + S_2 + S_3$ is less than the 1/4 area of the unit circle, we have

$$\sin x(\cos x - \cos y) + \sin y(\cos y - \cos z) + \sin z \cos z < \frac{\pi}{4}.$$

By the double angle formula, we have

$$\frac{\pi}{2} + 2\sin x \cos y + 2\sin y \cos z > \sin 2x + \sin 2y + \sin 2z.$$

Example 6. For any given positive integer n, find $2n + 1$ positive integers $a_i (1 \leq i \leq 2n + 1)$ such that they are an arithmetic sequence and their product is a perfect square.

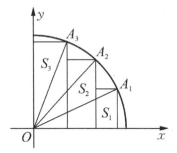

Figure 18.2

Solution. Take $a_i = ik (1 \leq i \leq 2n + 1)$, then $a_1 a_2 \ldots a_{2n+1} = (2n + 1)! k^{2n+1}$. Simply take $k = (2n + 1)!$ and the condition is met.

Remark. For any sequence $\{a_n\}$, let $a_1 a_2 \ldots a_{2n+1} = N$, and $(N a_1)(N a_2) \cdots (N a_{2n+1}) = N^{2n+2} = (N^{n+1})^2$ is a perfect square. And an arithmetic sequence times a number is still an arithmetic one. The reader is suggested to construct an arithmetic sequence of positive integers $\{a_n\}$, such that $a_1 a_2 \ldots a_{2n+1}$ has the form of $N^{2^m} (N \in \mathbb{N}^+)$, where m is a given positive integer.

It is a typical structural idea to construct an object that satisfies some properties and then modify it. When facing some large structural problems, it is often possible to decompose the big problem into several small ones and complete it step by step, so as to reduce the difficulty of construction. At the end of this chapter, Exercise 11 is based on the 6th Problem from the 2009 Chinese Mathematical Olympiad, in which each small problem is just a link in the constructive solution of the original problem.

Example 7. Find all positive integers n such that there exist non-zero integers $x_1, x_2, \ldots, x_n, y$, satisfying

$$\begin{cases} x_1 + \cdots + x_n = 0, \\ x_1^2 + \cdots + x_n^2 = ny^2. \end{cases} \tag{*}$$

(Problem from the 2007 Western China Mathematical Olympiad)

Solution. Obviously, $n \neq 1$.

If $n = 2k$, $k \in \mathbb{N}^+$, let $x_{2i-1} = 1$, $x_{2i} = -1$, $i = 1, 2, \ldots, k$, $y = 1$, and (*) holds.

If $n = 3 + 2k, k \in \mathbb{N}^+$, let $y = 2$, $x_1 = 4$, $x_2 = x_3 = x_4 = x_5 = -1$, $x_{2i} = 2$, $x_{2i+1} = -2$, $i = 3, 4, \ldots, k + 1$, and (*) holds.

If $n = 3$ and there exist non-zero integers x_1, x_2, x_3, such that

$$\begin{cases} x_1 + x_2 + x_3 = 0, \\ x_1^2 + x_2^2 + x_3^2 = 3y^2. \end{cases}$$

Eliminating x_3, we have $2(x_1^2 + x_2^2 + x_1 x_2) = 3y^2$.

Without loss of generality, we may assume that $(x_1, x_2) = 1$. Then $x_1^2 + x_2^2 + x_1 x_2$ is odd; on the other hand, $2 \mid y$, thus $3y^2 \equiv 0 \pmod 4$, but $2(x_1^2 + x_2^2 + x_1 x_2) \equiv 2 \pmod 4$, a contradiction.

To sum up, n can be any positive integer except 1 and 3.

Example 8. For positive integers $a_1, a_2, \ldots, a_{2006}$ such that a_1/a_2, $a_2/a_3, \ldots, a_{2005}/a_{2006}$ are distinct, find the minimum possible amount of distinct positive integers in the set $\{a_1, a_2, \ldots, a_{2006}\}$ (Problem from the 2006 China Mathematical Olympiad).

Solution. If there are n distinct values in $a_1, a_2, \ldots, a_{2006}$, then the number of distinct ratios in $a_1/a_2, a_2/a_3, \ldots, a_{2005}/a_{2006}$ is no more than $n(n-1)+1$. Thus, $n(n-1) + 1 \geq 2005$, that is, $n > 45$.

We construct an example such that $n = 46$ will work.

Let $p_1, p_2, \ldots, p_{46}$ be 46 distinct primes. Construct $a_1, a_2, \ldots, a_{2006}$ as follows:

$$p_1, p_1,$$

$$p_2, p_1,$$

$$p_3, p_2, p_3, p_1,$$

$$p_4, p_3, p_4, p_2, p_4, p_1,$$

$$\ldots$$

$$p_k, p_{k-1}, p_k, p_{k-2}, p_k, \ldots, p_k, p_2, p_k, p_1,$$

$$\ldots$$

$$p_{45}, p_{44}, p_{45}, p_{43}, p_{45}, \ldots, p_{45}, p_2, p_{45}, p_1,$$

$$p_{46}, p_{45}, p_{46}, p_{44}, p_{46}, \ldots, p_{46}, p_{34}.$$

It is easy to check that all the ratios are distinct. So, the minimum of n is 46.

Remark. This problem is a typical discrete extreme value problem, the solution is divided into two parts: one is to estimate the range of n, $n > 45$;

the second is to construct a concrete example to show that the extreme value in the range can be attained.

Extreme value problems in mathematical competitions have various backgrounds in algebra, geometry, and combinatorics. Oftentimes, one can use classic methods to estimate the range (in particular, lower or upper bound) of the variables; but to justify whether the extreme may be attained, usually one has to construct an example or counterexample. This situation is very common in discrete combinatorics problems, and an insightful example can even inspire the other part of the proof. We shall see the power of constructive method in the next two examples.

Example 9. Prove that for any positive integer m, we can find a finite set S of points in the plane, such that given any point $A \in S$, there are exactly m points in S at unit distance from A (Problem from 1971 International Mathematical Olympiad).

Solution. We prove by induction on m.

For $m = 1$, $S = S_1 = \{(0,0),(1,0)\}$ satisfies the property.

Suppose that for $m \in \mathbb{N}^+$, there exists a set $S = S_m$ that satisfies the property.

For $m + 1$, draw unit circles with centres at each point of S_m. Denote the intersections of these circles by T_m, which is finite. Then the directions of segments with any two endpoints in the set $S_m \cup T_m$ are finite. Take a unit direction $\vec{d}_m$ other than these directions, and let $S'_m = S_m + \vec{d}_m$. Now we show that $S = S_{m+1} = S_m \cup S'_m$. For any $A \in S_{m+1}$, by symmetry, we may assume $A \in S_m$, then by the induction hypothesis, there are exactly m points in S_m at unit distance from A, and there is exactly one point $A' = A + \vec{d}_m \in S'_m$ at unit distance from A. Otherwise, assume $B' = B + \vec{d}_m \in S'_m$ satisfies $AB' = 1$. Then $B' \in T_m$ and $\vec{d}_m$ is parallel to BB', a contradiction. Therefore, for all $m \in \mathbb{N}^+$, there exists a set $S = S_m$ that satisfies the property. This completes the proof.

Example 10. For arbitrary positive integer n, prove the following identity:

$$\sum_{k=0}^{n} 2^k \binom{n}{k}\binom{n-k}{\lfloor (n-k)/2 \rfloor} = \binom{2n+1}{n}.$$

(Problem of 1994 China Mathematical Olympiad)

Solution 1. (Construction of function)

Construct a function $f(x) = (1+x)^{2n+1}$: the coefficient of x^n in its expansion is exactly the right-hand side of the identity.

On the other hand,

$$f(x) = (1 + x)(1 + 2x + x^2)^n$$

$$= (1 + x) \sum_{k=0}^{n} \binom{n}{k} (1 + x^2)^{n-k} (2x)^k$$

$$= \sum_{k=0}^{n} 2^k \binom{n}{k} (1 + x)(1 + x^2)^{n-k} x^k.$$

If $n - k$ is even, the coefficient of x^{n-k} in $(1 + x)(1 + x^2)^{n-k}$ is $\binom{n-k}{(n-k)/2}$.

If $n - k$ is odd, the coefficient of x^{n-k} in $(1 + x)(1 + x^2)^{n-k}$ is $\binom{n-k}{(n-k-1)/2}$.

Thus, for $k = 0, 1, \ldots, n$, the coefficient of x^n in $2^k \binom{n}{k} (1 + x)(1 + x^2)^{n-k} x^k$ is always $2^k \binom{n}{k} \binom{n-k}{\lfloor (n-k)/2 \rfloor}$. This completes the proof.

Solution 2. (Construction of counting model)

Construct a model: there are n couples and a tourist guide, in which n persons visit place A and the other $n + 1$ persons visit place B.

On the one hand, there are $\binom{2n+1}{n}$ possible combinations.

On the other hand, assume that there are exactly $k(k = 0, 1, \ldots, n)$ couples visiting different places. There are $2^k \binom{n}{k}$ possible visiting plans for these k couples. And for the other $n - k$ couples and a tourist guide, if $n - k$ is even, then $(n - k)/2$ couples visit place A; if $n - k$ is odd, then $(n - k - 1)/2$ couples with the tourist guide visit place A; there are $\binom{n-k}{\lfloor (n-k)/2 \rfloor}$ possible plans in either case.

By the multiplication principle, if there are exactly $k(k = 0, 1, \ldots, n)$ couples visiting different places, there are a total of $2^k \binom{n}{k} \binom{n-k}{(n-k)/2}$ possible plans. Summing up for $k = 0, 1, \ldots, n$, we proved the identity.

Remark. In mathematical problem solving, it is often helpful to construct auxiliary functions or models: by investigating properties of the new function, one can retrieve some key features or solutions of the original problem (such as inequalities or extreme value problems); and by composing a model which maintains or realizes conditions and quantity relations of the original problem, one can consider a much simpler scenario and readily arrive at the conclusion. The model scenario usually comes from easy-to-understand imaginary situations in daily life.

The reader should practise the constructive method, so as to enhance comprehensive understanding of algebra, geometry, trigonometry, and other fields, and improve the analytical and problem-solving skills.

Exercises

1. Try to construct identities to prove the following Diophantine equations all have infinitely many positive integer solutions:

   (1) $x^2 + y^2 + 1 = z^2$;
   (2) $2x^2 + 2y^2 + 1 = z^2$;
   (3) $x^2 + y^2 + 1 = 2z^2$.

2. Find all positive real solutions of the system

$$\begin{cases} x^3 + y^3 + z^3 = x + y + z, \\ x^2 + y^2 + z^2 = xyz. \end{cases}$$

3. Minimize $f(x) = \sqrt{x^2 + 1} + \sqrt{(x-12)^2 + 16}$, $x \in \mathbb{R}$.

4. Let $x, y \in \left[-\frac{\pi}{4}, \frac{\pi}{4}\right]$, $a \in \mathbb{R}$, satisfy

$$\begin{cases} x^3 + \sin x - 2a = 0, \\ 4y^3 + \frac{1}{2} \sin 2y + a = 0. \end{cases}$$

   Find the value of $\cos(x + 2y)$.

5. Let real numbers α and β satisfy $\alpha^3 - 3\alpha^2 + 5\alpha = 1$, $\beta^3 - 3\beta^2 + 5\beta = 5$, find the value of $\alpha + \beta$.

6. Solve the system $\sum_{i=1}^{n} x_i^k = n (k = 1, 2, \ldots, n)$.

7. (a) Determine if the set $\{1, 2, \ldots, 96\}$ can be partitioned into 32 sets of equal sizes with equal sums.

   (b) Determine if the set $\{1, 2, \ldots, 99\}$ can be partitioned into 33 sets of equal sizes with equal sums.

   (Problem from the 2008 China Girls' Mathematical Olympiad)

8. Can we colour all positive integers with 2009 colours such that

   (1) Each colour paints infinitely many integers,
   (2) There are no three positive integers a, b, c in three different colours such that $a = bc$.

   (Problem from the 2009 All-Russian Mathematical Olympiad)

9. Show that there are infinitely many pairs of positive integers (m, n), such that $\frac{n+1}{m} + \frac{m+1}{n}$ is a positive integer (Problem from the 2007 British Mathematical Olympiad).

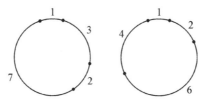

Figure 18.3

10. Given a circle with its perimeter equal to $n \in \mathbb{N}^+$, the least positive integer P_n which satisfies the following condition is called the *number of the partitioned circle*: there are P_n points $(A_1, A_2, \ldots A_{P_n})$ on the circle; for any integer $m (1 \le m \le n - 1)$, there always exist two points $A_i, A_j (1 \le i, j \le P_n)$, such that the length of arc $A_i A_j$ is equal to m. Furthermore, all arcs between every two adjacent points $A_i, A_{i+1} (1 \le i \le P_n, A_{P_n+1} = A_1)$ form a sequence $T_n = (a_1, a_2, \ldots a_{P_n})$ called the *sequence of the partitioned circle*. For example, when $n = 13$, the number of the partitioned circle $P_{13} = 4$, the sequence of the partitioned circle $T_{13} = (1, 3, 2, 7)$ or $(1, 2, 6, 4)$. Determine the values of P_{21} and P_{31}, and find the corresponding sequences T_{21} and T_{31}, respectively (Fig. 18.3) (Problem from the 2006 South East China Mathematical Olympiad).

11. Given an integer $n > 2$, let $f(X)$ denote the arithmetic mean of the finite set X of positive integers.

    (a) Prove that there exists a set S_1 with n positive integers, such that for any two different non-empty subsets A and B of S_1, $f(A)$ and $f(B)$ are two different integers.

    (b) If S_1 satisfies the property of (a), prove that for given positive integer $K > \max_{A_1 \in S_1} f(A_1)$ and any $x \in \mathbb{N}^+$, set $S_2 = \{K! x a + 1 | a \in S_1\}$ satisfies: for any two different non-empty subsets A and B of S_2, $f(A)$ and $f(B)$ are two coprime integers.

    (c) Prove that: based on (b), there exists an $x \in \mathbb{N}^+$, such that S_2 satisfies further that for any non-empty subset A of S_2, $f(A)$ is a composite number.

Chapter 19

Invariants and Monovariants

There are many invariants, that is, they do not change under movement, operation, and transformation. The method of invariants is to find some unchanging nature to solve the problem.

In the process of solving problems, we can find invariants in the form of sums, differences, products, quotients, sum of squares, sum of reciprocals, or signs, parites, congruences, and so on. Besides, a lot of colouring and assignment problems are actually solved by finding some sort of invariance in categorizing things or assigning values to things (cf. Chapters 14 and 15).

Example 1. Let $a_1, a_2, \ldots, a_9$ be non-zero real numbers. Show that, in the six terms of the determinant

$$\begin{vmatrix} a_1 & a_2 & a_3 \\ a_4 & a_5 & a_6 \\ a_7 & a_8 & a_9 \end{vmatrix} = a_1 a_5 a_9 + a_2 a_6 a_7 + a_3 a_4 a_8 - a_1 a_6 a_8 - a_2 a_4 a_9 - a_3 a_5 a_7,$$

there is at least one negative and one positive.

Solution. Since the sign of the product $-(a_1 a_2 \ldots a_9)^2$ of these six terms $a_1 a_5 a_9$, $a_2 a_6 a_7$, $a_3 a_4 a_8$, $-a_1 a_6 a_8$, $-a_2 a_4 a_9$ and $-a_3 a_5 a_7$ is negative, there must be an odd number of negative terms in the six terms. Thus, the result follows.

Example 2. There are 97 numbers $48/k$, $k = 1, 2, \ldots, 97$, on the blackboard. Each time, erase any two numbers a and b and write the number $2ab - a - b + 1$ instead on the blackboard. After 96 operations as above, what number is on the blackboard?

Solution. Since the number $2ab - a - b + 1 = 2(a - 1/2)(b - 1/2) + 1/2$ is equal to $1/2$ if a or $b = 1/2$, we see that $1/2$ is on the blackboard each time. Thus, the number left on the blackboard is $1/2$.

Example 3. A group of people are sitting around a round table debating on a problem with two opposing opinions. They express the opinions clockwise in turn. Each one approves of the opinion that most of his/her friends expressed. Prove that all people do not change their opinions after certain rounds of debates.

Solution. Consider the number of pairs of friends with opposing opinions. The number is finite. When one changes one's mind to follow the opinions of most of his/her friends, the number strictly decreases. So, one cannot change one's opinion infinitely. Thus, after certain rounds of debates, no person changes their mind.

Example 4. There are 2011 numbers $1, 2, \ldots, 2011$ on a line in the order from left to right. Prove that we cannot change the order of the numbers to $2011, 2010, \ldots, 2, 1$ from left to right by a finite sequence of operations each of which changes the contiguous three numbers (a, b, c) to (b, c, a).

Solution. For each permutation $P = (a_1, a_2, \ldots, a_{2011})$ of $(1, 2, \ldots, 2011)$, if $i < j, a_i > a_j$, we call (a_i, a_j) a pair of *inverse orders* of P. The number of such pairs is called *the number of inverse orders of P*, and denoted by $I(P)$. In the following, we shall prove that the parity of $I(P)$ is invariant under the operation. Let the permutation $P = (\ldots, a, b, c, \ldots)$ be changed to the permutation $P' = (\ldots, b, c, a, \ldots)$. If there are(is) k pair(s) of inverse orders in (a, b) and (a, c) of P, $k = 0, 1, 2$, then there are(is) $2 - k$ pair(s) of inverse orders in (b, a) and (c, a) of P'. Other pairs are inverse orders of P if and only if they are inverse orders of P'. Thus

$$I(P') - I(P) = (2 - k) - k \equiv 0 \pmod 2.$$

For $P_0 = (1, 2, \ldots, 2011)$ and $P_1 = (2011, 2010, \ldots, 1)$, we have

$$I(P_0) = 0, I(P_1) = \frac{2011 \times 2010}{2} \equiv 1 \pmod 2.$$

This completes the proof.

Remark. *Reverse order pair* and *inverse order number* are concepts often involved in permutation problems. Considering the change rules of inverse order number (such as monotonicity, parity, upper bound of each increment, etc.) is often the key to solving the problem. In this problem, we are using the *invariance* of the parity of the *inverse order number*.

One basic and important conclusion about inverse order number is as follows.

For n distinct real numbers $a_1, a_2, \ldots, a_n$, the parity of *inverse order number* changes when any two numbers exchange their positions.

Specially, by two successive swaps $(a, b) \rightarrow (b, a)$ and $(a, c) \rightarrow (c, a), (a, b, c)$ becomes (b, c, a), the *invariance of the parity* is obtained.

Example 5. Two sequences $\{a_n\}, \{b_n\}$ are defined by $a_1 = 1, b_1 = 2$, and

$$a_{n+1} = \frac{1 + a_n + a_n b_n}{b_n}, \quad b_{n+1} = \frac{1 + b_n + b_n a_n}{a_n}.$$

Show that $a_{2008} < 5$ (Problem from the 2008 All-Russian Mathematical Olympiad).

Solution. Since we know that

$$1 + a_{n+1} = \frac{(1 + a_n)(1 + b_n)}{b_n}, \quad 1 + b_{n+1} = \frac{(1 + b_n)(1 + a_n)}{a_n},$$

and $a_n, b_n > 0$ by the recurrence relations, thus, if $n \in \mathbb{N}^+$, then

$$\frac{1}{1 + a_{n+1}} - \frac{1}{1 + b_{n+1}} = \frac{b_n - a_n}{(1 + a_n)(1 + b_n)} = \frac{(1 + b_n) - (1 + a_n)}{(1 + a_n)(1 + b_n)}$$

$$= \frac{1}{1 + a_n} - \frac{1}{1 + b_n}$$

is an invariant. Thus,

$$\frac{1}{1 + a_{2008}} > \frac{1}{1 + a_{2008}} - \frac{1}{1 + b_{2008}} = \frac{1}{1 + a_1} - \frac{1}{1 + b_1} = \frac{1}{2} - \frac{1}{3} = \frac{1}{6},$$

which means $a_{2008} < 5$.

Example 6. Can you transform the quadratic trinomial $f(x) = x^2 + 4x + 3$ to $x^2 + 10x + 9$ by a sequence of transformations of $f(x)$ to $x^2 f(1 + 1/x)$ or to $(x - 1)^2 f(1/(x - 1))$?

Solution. The answer is negative.

The discriminant of $f(x) = ax^2 + bx + c$ is $\Delta = b^2 - 4ac$.

For the first transformation: $x^2 f(1 + 1/x) = (a + b + c)x^2 + (b + 2a)x + a$, its discriminant $\Delta_1 = (b + 2a)^2 - 4(a + b + c)a = b^2 - 4ac$.

For the second transformation: $(x - 1)^2 f(1/(x - 1)) = cx^2 + (b - 2c)x + (a - b + c)$, its discriminant $\Delta_2 = (b - 2c)^2 - 4c(a - b + c) = b^2 - 4ac$.

Thus, the discriminant is an invariant.

Since the discriminant of $x^2 + 4x + 3$ is $4^2 - 4 \times 3 = 4$, and that of $x^2 + 10x + 9$ is $10^2 - 4 \times 9 = 64$, the answer follows.

Example 7. In the following sequence:

$$1, \ 0, \ 1, \ 0, \ 1, \ 0, \ 3, \ldots,$$

the nth $(n > 6)$ term of the sequence is the last digit of the sum of the six terms before it. Prove that there are no successive six terms as 0, 1, 0, 1, 0, 1.

Solution. Let the successive six terms x, y, z, u, v, w of the sequence correspond to the number $2x + 4y + 6z + 8u + 10v + 12w$.

For example, the beginning six terms 1, 0, 1, 0, 1, 0, correspond to the number

$$2 \times 1 + 4 \times 0 + 6 \times 1 + 8 \times 0 + 10 \times 1 + 12 \times 0 = 18.$$

Let x, y, z, u, v, w, r be the successive seven terms, and B and A, the corresponding numbers of y, z, u, v, w, r and x, y, z, u, v, w, respectively. Then,

$$B - A = (2y + 4z + 6u + 8v + 10w + 12r)$$

$$- (2x + 4y + 6z + 8u + 10v + 12w)$$

$$= 12r - 2(x + y + z + u + v + w) \equiv 12r - 2r \equiv 0 \, (\mathrm{mod} \, 10).$$

Thus, each successive six terms of the sequence correspond to a number with the same last digit 8.

But six terms 0, 1, 0, 1, 0, 1 correspond to the number

$$2 \times 0 + 1 \times 1 + 6 \times 0 + 8 \times 1 + 10 \times 0 + 12 \times 1 = 24.$$

Its last digit is 4. This completes the proof.

Remark. The invariant is determined as follows: let x, y, z, u, v, w be the successive six terms of the sequence, consider the quantity $ax + by + cz + du + ev + fw$ with undetermined integers a, b, c, d, e, f such that if $r \equiv x + y + z + u + v + w \pmod{10}$, then

$$(ay + bz + cu + dv + ew + fr) - (ax + by + cz + du + ev + fw)$$

$$\equiv (f - a)x + (a + f - b)y + (b + f - c)z + (c + f - d)u$$

$$+ (d + f - e)v + ew \pmod{10}.$$

Choose a, b, c, d, e, f such that the coefficients of $x, y, z, u, v,$ and w are all $\equiv 0 \pmod{10}$.

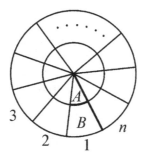

Figure 19.1

Example 8. Divide a circular pond into $2n(n \geq 5)$ parts as in Fig. 19.1. We call two parts *neighbours* if they have a common side or arc. Thus, every part has three *neighbours*. Now there are $4n + 1$ frogs at the pond. If there are three or more frogs in one part, then three frogs at this part will jump to their three *neighbourhoods*, respectively. Prove that some time later, the frogs at the pond will uniformly distribute. That is, for any part either there is at least one frog at it or there are frogs at each of the *neighbourhoods* (Problem from the 2005 China Mathematical Olympiad).

Solution. We call the event where three frogs jump to three neighbourhoods as *jumps*, and say a part is in *stable state* if there is at least one frog in it or its three neighbourhoods all have. Why do we use the words *stable state*? The reason is, once a part is in *stable state*, it remains in *stable state* forever.

It suffices to prove that for any sector with <3 frogs, the number will increase at some time. Since then every part is in *stable state* forever.

Without loss of generality, we prove it for sector 1 by contradiction. If the number of frogs at sector 1 is always less than 3 and never increases, then there are no *jumps* occurring at sectors 2 and n. Since $4n+1 > 4n-2$, by the drawer principle, at least one *jump* occurs at some sector k, $k = 3, \ldots, n-1$.

Define an *inertia* $\sum_{k=1}^{n} k^2 n_k$, where n_k is the number of frogs at sector k. Then if a *jump* occurs at sector $k(k = 3, \ldots, n-1)$, then the *inertia* increases by

$$k - 1^2 - 2k^2 + k + 1^2 = 2.$$

Since the *jumps* never stop, the *inertia* will increase to infinity, on the other hand, the *inertia* $\leq 4n + 1n^2$ by definition, a contridiction.

Exercises

1. Let an *operation* be as follows: replace any two numbers a and b in three numbers a, b, and c by $\frac{\sqrt{2}}{2}(a + b)$ and $\frac{\sqrt{2}}{2}(a - b)$, respectively. Can you obtain 3, 13, and 20 from 5, 12, and 18 by finite times of the *operations*?

2. There are 100 numbers $1, 1/2, \ldots, 1/100$ on the blackboard. Each time erase any two numbers a and b with the number $a + b + ab$ instead on the blackboard. After 99 times of above operations, what number is on the blackboard?

3. Draw a grid of 4×4 squares on the plane, and fill each of these squares arbitrarily with 1 or -1. The following changes of filling numbers is called an *operation*: take any square, calculate the product of all numbers in all its neighbouring squares (excluding the square itself). After taking all the squares, fill the calculated numbers into the corresponding squares. Is it possible to change the numbers of all squares to 1 after a finite number of *operations* for any initial fillings?

4. There are 45 chameleons living on an island, of which 13 are grey, 15 are brown, and 17 are purple. Whenever two chameleons in different colours meet, they both change to the third color. Can all 45 chameleons turn to the same colour some time later? (Problem from the Fall 1984 International Mathematics Tournament of Towns).

5. In a planar coordinate system, we got four pieces on positions with integer coordinates. You can make a move according to the following rule: you can move a piece to a new position if there is one of the other pieces in the middle of the old and new position. Initially, the four pieces have positions $\{(0,0), (0,1), (1,0), (1,1)\}$. Given a finite number of moves, can you yield the configuration $\{(0,0), (1,1), (3,0), (2,-1)\}$? (Problem from the 2008 Bundeswettbewerb Mathematik).

6. There is a grid of $n \times n$ squares, from which any $n - 1$ squares are allowed to be coloured black first, and step by step, the squares that are adjacent to at least two black squares are also coloured black. Prove that no matter how you first select the $n - 1$ squares, you cannot colour all of them black.

7. A positive integer is written on the blackboard. Whenever number x is written, one can write any one of the numbers $2x + 1$ and $x/(x + 2)$. At some moment, the number 2008 appears on the blackboard. Show that it was there from the very beginning (Problem from the 2008 All-Russian Mathematical Olympiad).

8. 2008 white stones and 1 black stone are in a row. An *action* means the following: select one black stone and change the colour of its neighbouring stone(s).

Find all possible initial positions of the black stone, such that one can make all stones black by finite *actions* (Problem from the 2009 Korea Mathematical Olympiad Final Round).

Chapter 20

Graph Theory Method

Graph theory stems from the famous Königsberg seven bridges problem, and studies the theory of the graph $G(V, E)$ which consists of vertices V and edges E, where E is a subset of pairs of V. The order of graph G is the number of vertices; it is denoted by $|G|$. And the size of the graph G is the number of edges; it is denoted by $e(G)$.

An edge of graph G is denoted by a pair of its two vertices, say v_1, v_2. If the edge is one way, we call it a directed edge, that is it has a direction from v_1 to v_2, denoted by an ordered pair of points (v_1, v_2), and draw the edge by a line segment between two points with an arrow when drawing the graph on the paper. If the edge is two way, it is also called an undirected edge, denoted by either (v_2, v_1) or (v_1, v_2), and we say that this edge joins the vertices v_1 and v_2 or vertices v_1, v_2 are incident with the edge (v_1, v_2). A graph G is called undirected if all its edges are undirected, otherwise it is called directed. A graph G is called oriented if each edge of G is directed.

In a graph G the position of vertices and the shape and the length of edges are irrelevant.

If (v_1, v_2) is a directed edge of graph G, then we say that v_1 is adjacent to v_2. If (v_1, v_2) is an undirected edge, then v_1 and v_2 are adjacent to each other, and we say that the vertices v_1 and v_2 are incident with the edge (v_1, v_2). If $v_1 = v_2$, then the edge (v_1, v_2) is called a loop, an edge joining a vertex to itself.

If several edges join the same two vertices, these edges are called multiple edges.

A graph G is called simple if G does not contain a loop or multiple edges.

A simple graph G is called complete if any two vertices of G are incident.

A complete graph of order n is denoted by K^n.

A graph with finite order and size is called a finite graph, otherwise it is called an infinite graph.

The out-degree of a vertex v of a graph G is the number of vertices that v is adjacent to. And the in-degree of a vertex is the number of vertices that are adjacent to v. For undirected graphs, the in-degree and out-degree of a vertex v are equal, and called the degree of the vertex v, denoted by $d(v)$. A vertex v is called odd if $d(v)$ is odd, and is called even if $d(v)$ is even.

In a graph G, a sequence consisting of vertices $\{v_0, v_1, \ldots, v_m\}$, where $e_i = (v_{i-1}, v_i)$, $i = 1, 2, \ldots, m$ are distinct edges of G, is called a trail from v_0 to v_m, where v_0 and v_m are called the initial vertex and the terminal vertex of the trail, and the number m is called the length of the trail. A trail that $v_0 = v_m$ is called a *circuit*. A *path* of the graph is a trail with distinct vertices. A cycle $\{v_0, v_1, \ldots, v_m\}$ is a circuit with distinct vertices $v_0, v_1, \ldots, v_{m-1}$.

A graph is connected if for every pair $\{u, v\}$ of distinct vertices there is a path from u to v or from v to u.

There is a theorem concerning the relation between the number of edges and the degrees of the vertices.

Theorem 1. *The sum of degrees of all vertices of an undirected graph G of order n is twice the number of edges, as follows:*

$$d(v_1) + d(v_2) + \cdots + d(v_n) = 2e(G).$$

Proof. Since each edge has two end vertices, the sum of the degrees is exactly twice the number of edges.

In particular, the sum of degrees is even, which sometimes is called the handshaking lemma, since it expresses the fact that in any party the total number of hands shaken is even. Thus, we have the following corollary.

Corollary. The number of odd vertices of an undirected graph is even.

One touch drawing is to trace each edge of an undirected graph G on the paper exactly once without lifting the pencil. If this is possible, then we say that graph G has a Eulerian trail; if it starts and ends at the same vertex, then we say that the graph G has a Eulerian circuit.

Theorem 2. *A connected undirected graph G has a Eulerian circuit if and only if each vertex is even. And G has a Eulerian trail but not circuit if and only if there are exactly two odd vertices, and these two vertices are the initial and terminal vertices.*

Most of the examples and exercises selected in this chapter are not graph theory problems, but we use graph theory method to solve them, aiming to reflect the extensive and flexible applications of graph theory.

Example 1. There are n tennis players who have scheduled exactly $n + 1$ two-person games among themselves. Show that at least one player plays at least three games.

Solution. In a simple graph G, n vertices $v_1, v_2, \ldots, v_n$ represent the n payers, and v_i, v_j are adjacent if and only if they played. Since there are $n + 1$ plays, the size of G is $n + 1$. By Theorem 1,

$$d(v_1) + d(v_2) + \cdots + d(v_n) = 2(n + 1).$$

On the other hand, if each player plays at most twice, we have

$$d(v_1) + d(v_2) + \cdots + d(v_n) \leq 2n,$$

which is a contradiction.

Example 2. Let $S = \{x_1, x_2, \ldots, x_n\}$, $n \geq 3$, be a set of points on the plane. If the distance between any two points of the set is not less than 1, show that no more than $3n$ pairs of points in the set have distance exactly 1.

Solution. Take these n points as vertices of the graph G, and two vertices are adjacent if and only if the distance between them is 1.

For each $i \in \{1, 2, \ldots, n\}$, the vertices of G adjacent to vertex x_i are located on the circle with radius 1 and centre x_i. Since the distance between any two points of the set S is not less than 1, there are at most six points of the set on the circle. Thus, the degree d_i of $x_i \leq 6$.

Let the size of G be e, then by Theorem 1, we have

$$2e = d_1 + d_2 + \cdots + d_n \leq 6n.$$

That is, $e \leq 3n$.

Remark. We can enhance the result $3n$ to $3n - 6$ by considering the convex hull of the set S.

Example 3. There are n lines in space such that in any three lines of them there are at least one pair but not all pairs of them on a plane. Find the maximal possible n.

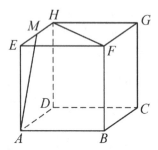

Figure 20.1

Solution. First, we show that $n \leq 5$.

If $n = 6$, denote six lines l_i by six vertices $v_i (1 \leq i \leq 6)$. If l_i and l_j are on a plane, (v_i, v_j) is red; if l_i and l_j are not on a plane, (v_i, v_j) is blue. We have a two coloured complete graph K^6. By Ramsey's theorem, in a two-coloured complete graph K^6 there is a monochromatic triangle. That is, there are three lines such that any two lines of them are on a plane or not on a plane, a contradiction. Thus, $n \leq 5$.

Let $n = 5$, in the cubic $ABCD-EFGH$ of Fig. 20.1, M is the midpoint of EH. It is easy to verify that five lines AB, BF, FH, HM, and MA meet the requirement of the problem.

Remark. The problem is too complicated if considered by space state of configuration of the line. After being translated to the language of the graph theorem, it turns to a Ramsey problem and it is easily solved (cf. Example 1 of Chapter 14).

Example 4 (Exercise 2 of Chapter 15). Let m points in $\triangle ABC$, together with points A, B, and C, form the vertices of several non-overlapping small triangles. Colour points A, B, and C in red, yellow, and blue, respectively, and colour each of the m points in $\triangle ABC$ in one of these three colours arbitrarily. Is the number of small triangles whose vertices are in distinct colours odd or even?

Solution. Construct a graph G as follows: take each small triangle and $\triangle ABC$ as the vertices of G. Vertices u, v of G are incident if triangles u and v have a common side with red and yellow vertices.

A triangle with three different coloured vertices corresponds to a vertex with degree 1, and other small triangles correspond to vertices with degree 0 or 2. By the corollary, the number of odd vertices is even, and $\triangle ABC$ corresponds to an odd vertex, so the answer is odd.

Remark. After being transformed into graph problems, many problems can be solved by existing theorems of graph (the previous Example 3 is a typical example of this aspect). Starting from the odd–even analysis, we simply use the well-known conclusion that 'the number of odd vertices is even' in a finite graph.

Example 5 (Exercise 3 of Chapter 16). In a school, there are b teachers and c students. Suppose that

(a) each teacher teaches exactly k students, and
(b) for any two (distinct) students, exactly h teachers teach both of them.

Prove that $b/h = c(c-1)/(k(k-1))$ (Problem from the 2004 Hong Kong Mathematical Olympiad).

Solution. Construct a graph G: G has two collections of vertices A and B, each teacher corresponds to a vertex of A and each pair of students corresponds to a vertex of B. Let $u \in A$, $v \in B$, then (u, v) is an edge if teacher u teaches both students of v.

By (a), each vertex of A is adjacent to $\binom{k}{2} = k(k-1)/2$ vertices of B. So, the size of G is $bk(k-1)/2$.

By (b), each vertex of B is adjacent to h vertices in A. So, the size of G is

$$h \binom{c}{2} = hc(c-1)/2.$$

Thus, $\frac{b}{h} = \frac{c(c-1)}{k(k-1)}$.

Remark. The proof of this problem uses the idea of modelling: construct a graph model to explain the problem conditions. In the two groups of vertices of graph G above, there are no adjacent vertices in each group, such a graph is called a *bipartite graph*. Starting from each group of vertices, we count the number of edges of the bipartite graph and get the desired equality of the two quantities.

Example 6. There are 34 couples dancing in a party. Before the dance, some dancers shake hands with other dancers but not with their own partners. Later, a male dancer asks the other 67 dancers their numbers of handshakes and he gets distinct answers. What is the number of his partner's handshakes?

Solution. Let each vertex of G represent a dancer. For vertices u and v, u and v are adjacent if they shake hands.

Since each couple does not shake hands themselves, the degree of any vertex is less than or equal to 66.

Let vertex x be the male dancer. The other 67 vertices have distinct degrees, and we denote the vertex with degree j by $x_j, j = 0, 1, \ldots, 66$. Since x_{66} shakes hands with all other 66 dancers except x_0, x_{66} and x_0 must be a couple.

Delete the couple x_{66} and x_0 and all edges that join the vertex x_{66}, and we have a graph G_1 with vertex x and 65 vertices with $d(x_j) = j - 1, j = 1, 2, \ldots, 65$. Discussing in the same way, we see that x_{65} and x_1 must be a couple. Continuing in this way, we can get that for $0 \le j \le 32, x_{66-j}$ and x_j must be a couple. Finally, the last couple are x_{33} and x. Thus, the number of handshakes of the partner x_{33} of the male dancer x is 33.

Remark. Numbering the vertices of a finite graph G in the decreasing or increasing order is one of the basic characteristics of the graph.

Example 7. For any n points $A_1, A_2, \ldots, A_n$ on a circle O, there are at least 2010 angles in $n(n-1)/2$ angles $\angle A_i O A_j (1 \le i < j \le n)$ no more than $120°$. Find the minimal n.

Solution. First, if $n = 90$, let AB be a diameter of circle O, take 45 points near point A and 45 points near point B as in Fig. 20.2. We see that only $2 \times 45 \times 44/2 = 45 \times 44 = 1980$ angles are no more than $120°·$

If $n = 91$, we show that there are at least 2010 angles no more than $120°$. Take 91 points $A_1, A_2, \ldots, A_{91}$ on the circle as vertices $v_1, v_2, \ldots, v_{91}$ of a graph G; if $\angle A_i O A_j > 120°$, then (v_i, v_j) is an edge. So, there are no triangles in G.

Suppose that there are e edges in graph G.

If $e = 0$, then there are $91 \times 90/2 = 4095 > 2010$ angles no more than $120°$.

If $e \ge 1$, without loss of generality, suppose (v_1, v_2) is an edge. Since there are no triangles in the graph, vertices $v_i (i = 3, 4, \ldots, 91)$ are adjacent

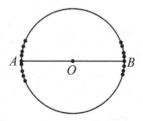

Figure 20.2

to at most one of v_1 and v_2, thus, $d(v_1) + d(v_2) \leq 89 + 2 = 91$. So, for any edge (v_i, v_j), $d(v_i) + d(v_j) \leq 91$.

Since $d(v_1) + d(v_2) + \cdots + d(v_{91}) = 2e$, summing up for each edge, we have

$$(d(v_1))^2 + (d(v_2))^2 + \cdots + (d(v_{91}))^2 \leq 91e.$$

By Cauchy's inequality,

$$91[(d(v_1))^2 + (d(v_2))^2 + \cdots + (d(v_{91}))^2] \geq [d(v_1) + d(v_2) + \cdots + d(v_{91})]^2 = 4e^2.$$

Therefore, $\frac{4e^2}{91} \leq (d(v_1))^2 + (d(v_2))^2 + \cdots + (d(v_{91}))^2 \leq 91e$, that is, $e \leq \frac{91^2}{4} < 2071$.

Hence, in 91 vertices, at least $91 \times 45 - 2070 = 2025$ pairs of points are not edges. So, the minimal number of n is 91.

For the following example, we shall use a directed graph in the solution.

Example 8. Let F be a finite set of integers, satisfying the following:

(a) for any $x \in F$, there are $y, z \in F$ (may be equal), such that $x = y + z$;
(b) there exists $n \in \mathbb{N}^+$, such that for any positive integer k $(1 \leq k \leq n)$ and any $x_1, x_2, \ldots, x_k \in F$ (may be equal), $\sum_{i=1}^{k} x_i \neq 0$.

Prove that F contains at least $2n+2$ elements (Problem from the Training for 2008 Chinese National Team of Mathematical Olympiad).

Solution. Obviously $0 \notin F$, and all elements of F are not of the same sign (otherwise element x with minimal absolute value does not satisfy (a).)

Let $x_1, x_2, \ldots, x_m$ be all positive elements of F. We show that $m \geq n + 1$.

Construct a directed graph G with m vertices $x_1, x_2, \ldots, x_m$ as follows: By (a), for each $i \in \{1, 2, \ldots, m\}$, there exist $x_j, z_k \in F$, such that $x_i = x_j + z_k$, $x_j > 0$, $i \neq j$, then let (x_i, x_j) be a directed edge of G.

Since each vertex x_i of G has out-degree 1 and the order of G is finite, there must be a directed circuit in graph G, say $x_{i_1} \to x_{i_2} \to \cdots \to x_{i_k} \to x_{i_1}$ $(k \leq m)$. This means there exist $z_1, z_2, \ldots, z_k \in F$, such that,

$$\begin{cases} x_{i_1} = x_{i_2} + z_1, \\ x_{i_2} = x_{i_3} + z_2, \\ \cdots \\ x_{i_k} = x_{i_1} + z_k. \end{cases}$$

Summing up the above equations, we have

$$z_1 + z_2 + \cdots + z_k = 0.$$

By (b), we must have $k > n$, thus $m \geq k \geq n + 1$.

Similarly, there are at least $n + 1$ negative integers in F. Hence, F has at least $2n + 2$ elements.

Remark. The result remains true when F is a finite set of real numbers. The reader can consider further whether F can contain exactly $2n + 2$ elements.

Exercises

1. Prove that for any group of two or more people, there are at least two persons having the same number of friends in the group.
2. Is there a polyhedron with an odd number of faces and each face has an odd number of edges?
3. Hundred people take part in a meeting. In any group of four persons, there is at least one who knows the other three persons in the group. At least how many people know all the other people in the meeting?
4. Prove that we can choose three numbers in any five irrational numbers such that the sum of any two of these three numbers is irrational.
5. Can you put 13 numbers $1, 2, \ldots, 13$ on a circle such that the absolute value of the difference of any two neighbouring numbers is at most 5 and at least 3?
6. If a convex n-gon has no three concurrent diagonals, into how many parts is the n-gon divided by its diagonals?
7. The 20 members of a local tennis club have scheduled exactly 14 two-person games among themselves, with each member playing in at least one game. Prove that within this schedule, there must be a set of 6 games with 12 distinct players (Problem from the 1989 American Mathematical Olympiad).
8. The sum of n positive integers $x_1, x_2, \ldots, x_n$ is 2009. If these n numbers can be divided into 41 groups of equal sum and can also be divided into 49 groups of equal sum, find the minimal of n (Problem from the 2009 Jiangxi preliminary competition for China National High School Mathematics League).
9. Ten distinct non-zero real numbers are given such that for any two, either their sum or their product is rational. Prove that squares of

all these numbers are rational (Problem from the 2005 All-Russian Mathematical Olympiad).

10. Let X be a set of $n(n \geq 2)$ elements. Define a 0-1valued function $f(x,y)$ on X by $f(x,y) + f(y,x) = 1(x \neq y, x, y \in X)$. Prove the following:

(a) The elements of X can be labeled as a sequence $\{x_1, x_2, \ldots, x_n\}$, such that $f(x_i, x_{i+1}) = 1$, for $i = 1, \ldots, n-1$. Moreover, for $n \geq 3$, define a *partition* of X $\{A, B\}$ by $X = A \cup B, A \neq \emptyset, B \neq \emptyset, A \cap B = \emptyset$. Then the sequence can be selected such that $f(x_n, x_1) = 1$ if and only if the following property holds.

(b) For any *partition* $\{A, B\}$ of X, there exists a pair $(x, y), x \in A, y \in B$ such that $f(x, y) = 1$.

Solutions for the Exercises

Chapter 1

1. **Solution 1.** Let $x = \sin\alpha, y = \sin\beta, \alpha, \beta \in \left[-\frac{\pi}{2}, \frac{\pi}{2}\right]$, then
$$x\sqrt{1-y^2} + y\sqrt{1-x^2} = \sin\alpha\cos\beta + \sin\beta\cos\alpha = \sin, (\alpha+\beta) \leq 1.$$
Equality holds when $\alpha + \beta = \frac{\pi}{2}$.
Solution 2. Sum up the following two inequalities by basic inequality:
$$x\sqrt{1-y^2} \leq \frac{1}{2}(x^2 + 1 - y^2), \quad y\sqrt{1-x^2} \leq \frac{1}{2}(y^2 + 1 - x^2).$$
Equalities hold when $x = y = \sqrt{2}/2$.

2. Taking the logarithm on both sides, we need only to prove $a\log b < b\log a$. To this end, we prove that $f(x) = \frac{\log x}{x}$ $(x > e)$ is decreasing, which is obvious since $f'(x) = \frac{1-\log x}{x^2} < 0$.

Remark. The above proof is perhaps the simplest. There are some other methods to prove the inequality, for example, use the method of *making the quotient*: let $f(x) = a^x/x^a$, then $f(a) = 1, f'(b) = f(b)(\ln(a)a/b) > 0$, for $b > a > e$. Thus, $f(b) > 1$, for $b > a > e$. We can also *make the difference*: let $f(x) = a^x - x^a$, then $f(a) = 0, f'(b) = a^b(\log(a) - a/b) + f(b)(a/b) > f(b)(a/b)$, for $b > a > e$. Since $(f(x)/x^a)'_{x=b} = [f'(b) - f(b)(a/b)]/b^a > 0$, thus, $f(b) > 0$, for $b > a > e$. However, this proof is much more complicated.

3. Write n as the sum of n of 1s:
$$n = 1 + 1 + \cdots + 1.$$
The problem is transformed to find the number of ways that can be used to select $(m-1) + s$ from $(n-1) + s$, which is $\binom{n-1}{m-1}$.

Remark. Counting problems are often 'simplified' by 'matching' elements in one set with elements in another set whose elements are easier to be counted.

4. Note that $x \geq 1, y \geq 1$, so,

$$(x^2 - 2x + 2)(y^2 - 2y + 2) - ((xy)^2 - 2xy + 2)$$
$$= (-2y + 2)x^2 + (6y - 2y^2 - 4)x + (2y^2 - 4y + 2)$$
$$= -2(y - 1)(x^2 + (y - 2)x + 1 - y)$$
$$= -2(y - 1)(x - 1)(x + y - 1) \leq 0.$$

Hence,

$$(x^2 - 2x + 2)(y^2 - 2y + 2) \leq (xy)^2 - 2xy + 2.$$

Similarly, since $xy \geq 1, z \geq 1$,

$$((xy)^2 - 2xy + 2)(z^2 - 2z + 2) \leq (xyz)^2 - 2xyz + 2.$$

5. If $a = 0$, we have

$$0 = (x_1 + x_2 + \cdots + x_n)^2 = \sum_{k=1}^{n} x_k^2 + 2 \sum_{1 \leq i < j \leq n} x_i x_j.$$

So, $\sum_{k=1}^{n} x_k^2 = -2 \sum_{1 \leq i < j \leq n} x_i x_j$. Thus, $\sum_{k=1}^{n} x_k^2 \leq 2 \sum_{1 \leq i < j \leq n} |x_i x_j|$. Hence,

$$2 \sum_{k=1}^{n} x_k^2 \leq \sum_{k=1}^{n} x_k^2 + 2 \sum_{1 \leq i < j \leq n} |x_i x_j| = \left(\sum_{k=1}^{n} |x_k| \right)^2.$$

That is, if $a = 0$, the inequality holds.

If $a \neq 0$, let $y_k = x_k - a, k = 1, 2, \ldots, n$, then the arithmetic mean of $y_1, y_2, \ldots, y_n$ is zero. Using the result proved above, we can get

$$\sum_{k=1}^{n} y_k^2 \leq \frac{1}{2} \left(\sum_{k=1}^{n} |y_k| \right)^2.$$

Hence, $\sum_{k=1}^{n} (x_k - a)^2 \leq \frac{1}{2} \left(\sum_{k=1}^{n} |x_k - a| \right)^2$.

Remark. The case of $a = 0$ is easy to prove. While for $a \neq 0$, the answer is obtained by transforming it into the case of $a = 0$. It is a commonly used method.

6. Let the lengths of BC, CA, and AB be a, b, and c, respectively, and the lengths of PD, PE, and PF be p, q, and r, respectively. We will find a point P inside $\triangle ABC$, such that $\frac{a}{p} + \frac{b}{q} + \frac{c}{r}$ attains its minimum. Since

$$ap + bq + cr = 2S_{\triangle PBC} + 2S_{\triangle PCA} + 2S_{\triangle PAB} = 2S_{\triangle ABC}$$

is a constant independent of the position of P,

By the AM-GM inequality $\frac{x}{y} + \frac{y}{x} \geq 2 (xy > 0)$, we have

$$(ap + bq + cr) \left(\frac{a}{p} + \frac{b}{q} + \frac{c}{r} \right) \geq (a + b + c)^2,$$

equality holds if and only if $p = q = r$. So, only when P is the incentre of $\triangle ABC$, $\frac{a}{p} + \frac{b}{q} + \frac{c}{r}$ attains its minimum $\frac{(a+b+c)^2}{2S_{\triangle ABC}}$.

7. (a) For given rational number a, we will find the rational numbers b and c, such that $(t + a)(bt + c) = 1$, that is,

$$bt^2 + (ab + c)t + ac - 1 = 0.$$

Change it to $(3b + ab + c)t - b + ac - 1 = 0$ by $t^2 = 3t - 1$. Since $t = (3 \pm \sqrt{5})/2$ are irrational numbers, we have

$$\begin{cases} 3b + ab + c = 0, \\ -b + ac - 1 = 0. \end{cases}$$

Solving for b and c, we have

$$b = -\frac{1}{a^2 + 3a + 1}, \quad c = \frac{a + 3}{a^2 + 3a + 1},$$

where $a^2 + 3a + 1 \neq 0$, since a is rational.

(b) Since $t^2 + 2 = (3t - 1) + 2 = 3t + 1 = 3 \left(t + \frac{1}{3} \right)$, and by (a), for $a = \frac{1}{3}$, we have

$$b = -\frac{1}{a^2 + 3a + 1} = -\frac{9}{19}, \quad c = \frac{a + 3}{a^2 + 3a + 1} = \frac{30}{19},$$

such that, $\left(t + \frac{1}{3} \right) \left(-\frac{9}{19}t + \frac{30}{19} \right) = 1$.
Therefore, $\frac{1}{t^2 + 2} = \frac{1}{3(t + 1/3)} = -\frac{3}{19}t + \frac{10}{19}$.

Remark. We simplify Problem (b) to a solved familiar Problem (a).

8. Let there be r of (-1)'s s of 1's, and t of 2's in $x_1, x_2, \ldots, x_n$. The equations

$$\begin{cases} -r + s + 2t = 19, \\ r + s + 4t = 99 \end{cases}$$

yield

$$\begin{cases} r = 40 - t, \\ s = 59 - 3. \end{cases}$$

So,

$$\begin{cases} r = 40 - t \geq 0, \\ s = 59 - 3t \geq 0, \\ t \geq 0. \end{cases}$$

Thus,

$$0 \leq t \leq 19.$$

$$x_1^3 + x_2^3 + \cdots + x_n^3 = -r + s + 8t = (19 - 2t) + 8t = 6t + 19.$$

Hence,

$$19 \leq x_1^3 + x_2^3 + \cdots + x_n^3 \leq 133.$$

If $r = 40$, $s = 59$, and $t = 0$, $x_1^3 + x_2^3 + \cdots + x_n^3 = 19$; if $r = 21$, $s = 2$, and $t = 19$, $x_1^3 + x_2^3 + \cdots + x_n^3 = 133$. So, the minimum and the maximum of $x_1^3 + x_2^3 + \cdots + x_n^3$ are 19 and 133, respectively.

9. Let $x + 2y = 5a$, $x + y = 3b$, then $x = 6b - 5a$, $y = 5a - 3b$. Consequently, $2x + y = 9b - 5a$, and

$$\begin{cases} 5a - 3b \geq 0, \\ 9b - 5a \geq 99. \end{cases}$$

Thus, $9b \geq 5a + 99 \geq 3b + 99$, which yields $b \geq 17$, then $5a \geq 3b \geq 51$, and this yields, $a \geq 11$, then $9b \geq 5a + 99 \geq 5 \cdot 11 + 99 = 154$, which yields, $b \geq 18$.

If $b = 18$, then $5a \leq 9b - 99 = 63$, $a \leq 12$, consequently,

$$7x + 5y = 27b - 10a \geq 27 \times 18 - 10 \times 12 = 366.$$

If $b > 18$, then $7x + 5y = 27b - 10a = 9b + 2(9b - 5a) \geq 9 \times 19 + 2 \times 99 = 369 > 366$.

Summing up, the minimum of $7x + 5y$ is 366, which is attained at $x = 48$ and $y = 6$.

10. Since $a + c = (1 - ac)b$, clearly $1 - ac \neq 0$. So, $b = \frac{a+c}{1-ac}$, which reminds us of the tangent formula for the sum of two angles. Let $\alpha = \arctan a$, $\beta = \arctan b$, $\gamma = \arctan c$, $\alpha, \beta, \gamma \in (0, \pi/2)$, then

$$\tan \beta = \frac{\tan \alpha + \tan \gamma}{1 - \tan \alpha \tan \gamma} = \tan(\alpha + \gamma).$$

Since $\beta, \alpha + \gamma \in (0, \pi)$, we have $\beta = \alpha + \gamma$. Thus,

$$P = \frac{2}{\tan^2 \alpha + 1} - \frac{2}{\tan^2 \beta + 1} + \frac{3}{\tan^2 \gamma + 1}$$

$$= 2\cos^2 \alpha - 2\cos^2(\alpha + \gamma) + 3\cos^2 \gamma$$

$$= (\cos 2\alpha + 1) - (\cos(2\alpha + 2\gamma) + 1 + 3\cos^2 \gamma$$

$$= 2\sin \gamma \cdot \sin(2\alpha + \gamma) + 3\cos^2 \gamma$$

$$\leq 2\sin \gamma + 3(1 - \sin^2 \gamma)$$

$$= -3(\sin \gamma - 1/3)^2 + 10/3 \leq 10/3.$$

The equality holds if $2\alpha + \gamma = \pi/2, \sin \gamma = 1/3$, that is, $a = \sqrt{2}/2$, $b = \sqrt{2}$, and $c = \sqrt{2}/4$. Therefore, the maximum of P is $10/3$.

Chapter 2

1. Suppose on the contrary that $|x| < |y-z|, |y| < |z-x|, |z| < |x-y|$ and $x \geq y \geq z$ might as well, then

$$|x| + z| < |y - z| + |x - y| = (y - z) + (x - y) = x - z.$$

A contradiction.

2. Suppose, on the contrary, that the length of each side of the quadrilateral $EFGH$ is less than $\sqrt{2}/2$. There is an angle, say, $\angle EFG \leq 90°$. Then $EG^2 \leq EF^2 + FG^2 < 1$. On the other hand, the points E and G are on two parallel sides AB and CD, respectively. So, $EG \geq 1$. A contradiction.

3. (a) The answer is negative. Suppose, on the contrary, there exist positive integers m and n, such that $m(m+2) = n(n+1)$, then $(m+1)^2 = n^2 + n + 1$.

Since

$$n^2 < n^2 + n + 1 < (n+1)^2,$$

we see that $n^2 + n + 1$ is not a perfect square. A contradiction.

(b) For $k = 3$, if there exist positive integers m and n, satisfying $m(m+3) = n(n+1)$, then

$$4m^2 + 12m = 4n^2 + 4n,$$

$$(2m+3)^2 = (2n+1)^2 + 8,$$

$$(2m+3 - 2n - 1)(2m+3+2n+1) = 8,$$

$$(m - n + 1)(m + n + 2) = 2.$$

This cannot hold since $m + n + 2 > 2$.

For $k = 2t$ (t is an integer greater than 1), take

$$m = t^2 - t, \quad n = t^2 - 1.$$

Then

$$m(m + k) = (t^2 - t)(t^2 + t) = t^4 - t^2,$$
$$n(n + 1) = (t^2 - 1)t^2 = t^4 - t^2 = m(m + k).$$

For $k = 2t + 1$ (t is an integer greater than 1), take

$$m = \frac{t^2 - t}{2}, \quad n = \frac{t^2 + t - 2}{2}.$$

Then

$$m(m + k) = \frac{t^2-t}{2}\left(\frac{t^2-t}{2} + 2t + 1\right) = \tfrac{1}{4}(t^4 + 2t^3 - t^2 - 2t),$$
$$n(n + 1) = \frac{t^2+t-2}{2} \cdot \frac{t^2+t}{2} = \tfrac{1}{4}(t^4 + 2t^3 - t^2 - 2t) = m(m + k).$$

Summing up, the answer is negative when $k = 3$, and is positive when $k \geq 4$.

4. Suppose, on the contrary, there exist three distinct integers x_1, x_2, x_3, such that

$$|ax_i^2 + bx_i + c| \leq 1000.$$

Let $f(x) = ax^2 + bx + c$, then at least two of x_1, x_2, x_3 are on one side of the symmetric axis $x = -\frac{b}{2a}$ (the axis is included), say, they are x_1, x_2, and

$$|x_1 + b/(2a)| \geq |x_2 + b/(2a)| + 1.$$

Thus,

$$|f(x_1) - f(x_2)| = |x_1 - x_2||a(x_1 + x_2) + b| \leq |f(x_1)| + |f(x_2)| \leq 2000.$$

$$(1)$$

On the other hand,

$$|x_1 - x_2||a(x_1 + x_2) + b| = |x_1 - x_2|(a|x_1 + b/(2a)| + a|x_2 + b/(2a)|)$$
$$\geq |x_1 - x_2|(a + 2a|x_2 + b/2a|) \geq a > 2000.$$

This contradicts (1).

5. The answer is negative. Suppose, on the contrary, there exists a ΔABC with $\angle A \leq \angle B \leq \angle C$, that $\angle C = 2\angle A$, and $a = 2007$. Let CD be the bisector of $\angle C$, then $\angle BCD = \angle A$, so $\Delta CDB \backsim \Delta ACB$. Hence,

$$\frac{CB}{AB} = \frac{BD}{BC} = \frac{CD}{AC} = \frac{BD+CD}{BC+AC} = \frac{BD+AD}{BC+AC} = \frac{AB}{BC+AC}.$$

That is, $c^2 = a(a+b) = 2007(2007+b)$, where $2007 \leq b \leq c < 2007+b$.

Therefore, $2007|c^2$, so $3|c$ and $223|c$. Let $c = 669m$, then $223m^2 = 2007 + b$, together with $b \geq 2007$, yields $m \geq 5$.

On the other hand, by $c \geq b$, we have, $669m \geq 223m^2 - 2007$, which means $m < 5$. A contradiction.

6. The answer is negative. Suppose, on the contrary, there exist infinitely many numbers $a_1, a_2, a_3, \ldots$ in A, such that for any $i, j \in \mathbb{N}^+$, $\frac{a_i}{i} + \frac{a_j}{j} \in A$. Obviously $a_1 \neq 1$(otherwise, let $i = j = 1$, then $2 \in A$).

Let $a_1 = \frac{1}{m}, m \geq 2, m \in \mathbb{N}^+, i = 1, j = m^2$, then

$$\frac{a_i}{i} + \frac{a_j}{j} = \frac{1}{m} + \frac{a_{m^2}}{m^2} \leq \frac{1}{m} + \frac{1}{m^2} < \frac{1}{m-1}.$$

That is $\frac{1}{m} < \frac{a_i}{i} + \frac{a_j}{j} < \frac{1}{m-1}$, and it contradicts $\frac{a_i}{i} + \frac{a_j}{j} \in A$.

7. Suppose, on the contrary, for any two three-member committees A and B, either they do not share any member, or they share exactly two members. For the latter, denote $A \sim B$.

First, we show that, if $A \sim B, B \sim C$, then $A \sim C$. In fact, let $A = \{a, b, c\}$, $B = \{a, b, d\}$. Since $B \sim C$, $C \cap \{a, b\} \neq \emptyset$. Hence, $A \sim C$ by the hypothesis.

Thus, all three-member committees can be divided into several classes by the relation $\sim$, so that any two three-member committees in the same class share exactly two members, and any two three-member committees in different classes do not share any member.

The number of members in each class is in one of the following cases.

(I) 3.

(II) 4.

(III) ≥ 5.

For case (I), the class consists of exactly one three-member committee.

For case (II), the class consists of at most 4 three-member committees.

For case (III), let $A = \{a, b, c\}, B = \{a, b, d\}$ be in the class. Then there exists C, which contains element e other than a, b, c, d. Since $A \sim C, B \sim C$, we see that $C = \{a, b, e\}$. For any other D in the class, by

$A \sim D, B \sim D, C \sim D$, we see that $\{a, b\} \subseteq D$. Thus, the number of committees in the class is 2 less than the members in the class.

On the other hand, the total number of committees $n + 1$ is greater than the number of total members n. A contradiction.

8. Suppose, on the contrary, there exists an i_0, such that for any $i > i_0, (a_i, a_{i+1}) > \frac{3}{4}i$. On the other hand, for a given positive integer $M(M > i_0)$, if $i \geq 4M$, then $(a_i, a_{i+1}) > \frac{3}{4}i \geq 3M$. So, $a_i \geq (a_i, a_{i+1}) > 3M$. We have, $\{1, 2, \ldots, 3M\} \subseteq \{a_1, a_2, \ldots, a_{4M-1}\}$. Consequently,

$$|\{1, 2, \ldots, 3M\} \cap \{a_{2M}, a_{2M+1}, \ldots, a_{4M-1}\}|$$
$$\geq 3M - (2M - 1) = M + 1.$$

By the drawer principle, there exists a $j_0 (i_0 < 2M \leq j_0 < 4M - 1)$, such that $a_{j_0}, a_{j_0+1} \leq 3M$.

Hence,

$$(a_{j_0}, a_{j_0+1}) \leq \frac{1}{2}\max\{a_{j_0}, a_{j_0+1}\} \leq \frac{3M}{2} = \frac{3}{4} \times 2M \leq \frac{3}{4}j_0.$$

A contradiction.

Chapter 3

1. Induction on n.

   If $n = 1$, $a_1^3 = a_1^2$ implies $a_1 = 1$, so the conclusion is true when $n = 1$.

   Suppose that the conclusion is true for $n \leq k$, that is, $a_i = i$ for $i = 1, 2, \ldots, k$. Then, for $n = k + 1$,

   $$a_{k+1}^3 + \sum_{i=1}^{k} i^3 = \left(a_{k+1} + \sum_{i=1}^{k} i \right)^2.$$

   So, $a_{k+1}(a_{k+1}^2 - a_{k+1} - k(k+1)) = 0$, which yields, $a_{k+1} = k + 1$, since $a_{k+1} > 0$.

2. In the following, we prove an enhanced result $a_{n+1} > 3b_n$ by induction on n.

   For $n = 1$, $a_2 = 3^3 = 27 > 24 = 3b_1$, that is, the conclusion is true for $n = 1$.

   Suppose that the conclusion is true for $n \geq 1$, that is, $a_{n+1} > 3b_n$. Then,

   $$a_{n+2} = 3^{a_{n+1}} > 3^{3b_n} = 27^{b_n} > 3^{b_n} \cdot 8^{b_n} > 3b_{n+1}.$$

   That is, $a_{n+2} > 3b_{n+1}$.

3. Since a square can be divided equally into four small squares, it's easy to increase the number of small squares by three. So, we prove by the third form of induction with induction base $n_0 = 6$ and induction step 3. For $n = 6, 7, 8$, the segmentation can be done as shown in Fig. 3(Ex).1.

   Suppose that a square can be divided into $n = k \geq 6$ small squares. Then dividing one square into four smaller squares, we obtain $n = k + 3$

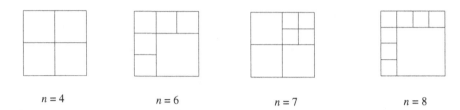

$n = 4$ $n = 6$ $n = 7$ $n = 8$

Figure 3(Ex).1

smaller squares. Thus, a square can be divided into any $n(\geq 6)$ small squares.

4. Induction on n.

(1) The conclusion is obviously true for $n = 1$.

(2) Suppose that the conclusion is true for $n = k \geq 1$. Without loss of generality, assume that, $n_1 > n_2 > \cdots > n_k > n_{k+1}$, we have,

$$2^{\frac{n_2}{2}} + 2^{\frac{n_3}{2}} + \cdots + 2^{\frac{n_{k+1}}{2}} < (1 + \sqrt{2})\sqrt{M - 2^{n_1}}.$$

Then,

$$2^{\frac{n_1}{2}} + 2^{\frac{n_2}{2}} + \cdots + 2^{\frac{n_{k+1}}{2}} < (1 + \sqrt{2})\sqrt{M - 2^{n_1}} + 2^{\frac{n_1}{2}}.$$

It suffices to show that

$$(1 + \sqrt{2})\sqrt{M - 2^{n_1}} + 2^{\frac{n_1}{2}} < (1 + \sqrt{2})\sqrt{M}. \tag{1}$$

Since $M = 2^{n_1} + 2^{n_2} + \cdots + 2^{n_{k+1}} \leq 2^{n_1} + 2^{n_1-1} + \cdots + 2^{n_1-n} < 2 \times 2^{n_1}$, $M - 2^{n_1} < 2^{n_1}$, hence,

$$\sqrt{M - 2^{n_1}} + \sqrt{M} < (1 + \sqrt{2}) \cdot 2^{\frac{n_1}{2}}.$$

Multiply both sides by $\sqrt{M} - \sqrt{M - 2^{n_1}}$ to yield

$$M - (M - 2^{n_1}) < (1 + \sqrt{2}) \cdot 2^{\frac{n_1}{2}} \sqrt{M} - \sqrt{M - 2^{n_1}},$$

which is equivalent to (1). That is, the conclusion is true for $k + 1$. Therefore, the conclusion is true for all positive integers.

5. First we prove that $a_n = n$ for $n = 2^k$, $k \in \mathbb{N}^+$ by induction on k.

(1) For $k = 1$, $a_2 = 2$.

(2) Suppose that $a_{2^k} = 2^k$ for $k \geq 1$, then $a_{2^{k+1}} = a_2 a_{2^k} = 2 \times 2^k = 2^{k+1}$.

So, for all $k \geq 1$, $a_{2^k} = 2^k$.

Now, we prove by reverse induction on n.

(1) Since we have just proved that $a_n = n$ for $n = 2^k$, $k \in \mathbb{N}^+$.

(2) Since $\{a_k\}$ is a strictly increasing sequence of positive integers, $a_1 = 1$, $a_2 = 2$, $a_n = n$. Thus, $a_k = k$ for all $1 \leq k < n$, and we see that such $\{a_k\}$ satisfies all conditions of the problem.

So, $a_n = n$ for all positive integers n.

6. We prove by the third form of induction with induction base $n_0 = 3$ and induction step $k = 2$.

(1) For $n = 3$ and 4, we see that $3^3 + 4^3 + 5^3 = 6^3$ and $1^3 + 5^3 + 7^3 + 12^3 = 13^3$ or $11^3 + 12^3 + 13^3 + 14^3 = 20^3$.

(2) Suppose that for $n \geq n_0 = 3$, we have $\sum_{i=1}^{n} a_i^3 = b^3$ with $a_1 < a_2 < \cdots < a_n$.

Then, $n + k = n + 2$ numbers $x_1 = 3a_1$, $x_2 = 4a_1$, $x_3 = 5a_1$, $x_4 = 6a_2$, $x_5 = 6a_3, \ldots, x_{n+2} = 6a_n$, and $y = 6b$ satisfy the equation $\sum_{i=1}^{n+2} x_i^3 = y^3$.

7. We prove by induction on n, that there are infinitely many positive integers $n_1, n_2, \ldots$, such that the elements of the set $A = \{2^{n_i} - 3 | i = 1, 2, \ldots\}$ are pairwise coprime.

(1) For $n = 2$, $2^2 - 3 = 1$ and $2^3 - 3 = 5$ are coprime.

(2) Suppose that $A_n = \{2^{n_i} - 3 | i = 1, 2, \ldots, n\}$ satisfies the property for n: any two elements of A_n are coprime for $n \geq 2$.

Let $p_1, p_2, \ldots, p_r$ be all the prime factors of elements of A_n. By the Fermat little theorem, $2^{p_j - 1} \equiv 1 \pmod{p_j}$, $j = 1, 2 \ldots, r$, hence, $2^{(p_1 - 1)(p_2 - 1) \cdots (p_r - 1)} \equiv 1 \pmod{p_j}$, for $j = 1, 2, \ldots, r$.

Let $n_{n+1} = (p_1 - 1)(p_2 - 1) \cdots (p_r - 1) \geq 2$, then for any $p_j, j \in \{1, 2, \ldots, r\}$, $2^{n_{n+1}} - 3 \equiv 1 - 3 \equiv -2 \pmod{p_j}$. Hence, $2^{n_{n+1}} - 3$ is not divisible by any p_j, $j \in \{1, 2, \ldots, r\}$, that is, the set $A_{n+1} = \{2^{n_i} - 3 | i = 1, 2, \ldots, n + 1\}$ satisfies the property in (2) for $n + 1$.

8. We prove that the number of ways is $(2n - 1)!!$ by induction on n.

(1) For $n = 1$, there is only one weight and it can only be put on the left pan. So, the number of ways is $(2n - 1)!! = 1$ for $n = 1$.

(2) Suppose that for $n \geq 1$ weights, the answer is $(2n - 1)!!$.

Then for $n + 1$ weights, multiply each weight by $1/2$ and this does not affect the result. Now the n weights are of weight $1/2, 1, 2, \ldots, 2^n$.

Since $2^r > \sum_{i=-1}^{r-1} 2^i \geq \sum_{i=-1}^{r-1} \pm 2^i$ for $r \in \mathbb{N}^+$, the position of the heaviest weight on the balance determines the heavier side of the balance. Therefore, the heaviest weight is on the left pan. Now consider the position of the weight of $1/2$ in the process of operations.

(a) If the weight of $1/2$ goes first, it can only be put on the left pan. Then the rest n weights have $(2n-1)!!$ ways.

(b) If the weight of $1/2$ goes at the tth operation, $t = 2, 3, \ldots, n+1$. Since the weight $1/2$ is not the heaviest, it can be put either on the left or right pan and does not change which side is heavier. So, there are $2 \times (2n-1)!!$ ways.

Summing up, there are totally $(2n-1)!! + n \times 2 \times (2n-1)!! = ((2(n+1)-1)!!$ ways of putting $n+1$ weights.

Therefore, the number of ways is $(2n-1)!!$ for n weights.

9. First, we see that $1 \in A$. Let $a \in A$, $b \in A$, $1 < a < b$. If at least one of a and b is even, then $2 \in A$; if a and b are all odd, then $1 + ab \in A$, and $1 + ab$ is even, so $2 \in A$.

If $1, 2, a \in A \, (a > 2)$, then $1 + 2 \cdot a \in A$, $1 + 2 \cdot (1 + 2a) = 3 + 4a \in A$,

$$1 + (1 + 2a) \cdot (3 + 4a) = 4 + 10a + 8a^2 \in A.$$

If a is even, then $4 \,|\, (4 + 10a + 8a^2)$, thus $4 \in A$; if a is odd, then take $4 + 10a + 8a^2$ as a and repeat the above procedure, to yield $4 \in A$.

Further, $1 + 2 \times 4 = 9 \in A$, so $3 \in A$. And $1 + 2 \times 3 = 7 \in A$, $1 + 2 \times 7 = 15 \in A$, so $5 \in A$.

Thus, $1, 2, 3, 4, 5$ are all elements of A.

Now we prove by induction on n.

(1) $n = 1, 2, 3, 4, 5$ are all elements of A.

(2) Suppose that $n \in A$, $(n \geq 5)$.

If $n + 1 = 2k + 1$, then $3 \leq k < n$, thus, $n + 1 = 1 + 2 \cdot k \in A$;

If $n + 1 = 2k$, then $3 \leq k < n$. Since $n = 2k - 1 \in A$ and $1 + 2 \cdot k \in A$, we have $1 + (2k-1) \cdot (2k+1) = 4k^2 \in A$, thus, $2k \in A$, that is, $n+1 \in A$.

Summing up, we have $A = \mathbb{N}^+$.

10. We enhance the conclusion to

$$AP_1^2 + P_1 P_2^2 + P_2 P_3^2 + \cdots + P_{n-1} P_n^2 + P_n B^2 \leq AB^2.$$

(1) For $n = 1$, since $\angle AP_1 B \geq 90°$, $AP_1^2 + P_1 B^2 \leq AB^2$. The conclusion is true for $n = 1$.

(2) Suppose that the conclusion is true for $n \geq 1$.

For $n + 1$ points, make the altitude CD of right-angled $\triangle ABC$ on the hypotenuse AB, so that there are s and t points in two right-angled triangles $\triangle ACD$ and $\triangle BCD$, respectively. (Otherwise, if $n + 1$ points are all in one small right-angled triangle, we can perform the same procedure on this small right-angled triangle, until these $n+1$ points are not all in one small right-angled triangle.) Then $s, t \geq 1$, $s + t = n + 1$.

By (2), we can label the points in $\triangle ACD$ with $P_1, P_2, \ldots, P_s$, such that, by the induction hypothesis,

$$AP_1^2 + P_1 P_2^2 + \cdots + P_{s-1} P_s^2 + P_s C^2 \leq AC^2.$$

And label the points in BCD with $P_{s+1}, P_{s+2}, \ldots, P_{n+1}$, such that

$$CP_{s+1}^2 + P_{s+1} P_{s+2}^2 + \cdots + P_n P_{n+1}^2 + P_{n+1} B^2 \leq BC^2.$$

Since $\angle P_s C P_{s+1} \leq 90°$, $P_s P_{s+1}^2 \leq P_s C^2 + C P_{s+1}^2$. Thus,

$$AP_1^2 + P_1 P_2^2 + \cdots + P_n P_{n+1}^2 + P_{n+1} B^2$$
$$\leq (AP_1^2 + P_1 P_2^2 + \cdots + P_s C^2)$$
$$+ (C P_{s+1}^2 + P_{s+1} P_{s+2}^2 + \cdots + P_n P_{n+1}^2 + P_{n+1} B^2)$$
$$\leq AC^2 + BC^2 = AB^2.$$

Hence, the conclusion is true for $n + 1$.

11. We prove by induction on n for n subjects, 2^n students and 2^{n-1} rooms.

   (1) For $n = 2$, the conclusion is true.
   (2) Suppose that the conclusion is true for $n - 1$. $n \geq 3$.

For $n \geq 3$, consider a student selects a subject, say, physics, and his/her roommate does not. Denote set A the students who select physics, and B who do not. Then $|A| = |B| = 2^{n-1}$. Let students of A and B be arranged in two schools with each having 2^{n-2} rooms, such that old roommates are still roommates, and the others are assigned arbitrarily to be the new roommates. By (2), the students of A can queue up in a circle K satisfying the requirements. Let $(x_1, x_2), \ldots, (x_{2k-1}, x_{2k})$ be the new roommates of A on circle K clockwise. Let x_i' be the old roommate of x_i, obviously $x_i' \in B$. In B, let $(x_2', x_3'), \ldots, (x_{2k-2}', x_{2k-1}')$, (x_{2k}', x_1') be the pairs of new roommates. By (2), the students of B can queue up in a circle K' satisfying the requirements. Then, place all students on K between x_{2i} and x_{2i+1} in the space between x_{2i}' and x_{2i+1}' on K' in order. The new larger circle satisfies the requirements.

Chapter 4

1. Divide $2n$ positive integers $1, 2, \ldots, 2n$ into n groups as follows:

$$\{1, 2\}, \{3, 4\}, \ldots, \{2n - 1, 2n\}.$$

Take $n + 1$ numbers from these groups. By the drawer principle, there are two numbers in the same group, which are coprime since they are consecutive.

2. Divide 100 numbers $1, 2, \ldots, 100$ into 10 groups as follows, such that the ratio of maximum and the minimum in each group does not exceed $3/2$.

$$\{1\}, \{2, 3\}, \{4, 5, 6\}, \{7, 8, 9, 10\}, \{11, 12, \ldots, 16\}, \{17, 18, \ldots, 25\},$$

$$\{26, 27, \ldots, 39\}, \{40, 41, \ldots, 60\}, \{61, 62, \ldots, 91\}, \{92, 93, \ldots, 100\}.$$

Take 11 numbers from $1, 2, \ldots, 100$, that is, take 11 numbers from 10 groups. By the drawer principle, there are 2 numbers a and b, $a > b$, in the same group, so, the ratio a/b does not exceed $3/2$.

3. Let the seven real numbers be $\tan \alpha_i$, $i = 1, 2, \ldots, 7$, where

$$-\frac{\pi}{2} < \alpha_1 \leq \alpha_2 \leq \cdots \leq \alpha_7 < \frac{\pi}{2}.$$

Divide the interval $[\alpha_1, \alpha_7]$ equally into six subintervals. By the drawer principle, there are two angles in $\alpha_i (1 \leq i \leq 7)$, say, $\theta_1, \theta_2 (\theta_1 \leq \theta_2)$ in the same subinterval, then

$$0 \leq \theta_2 - \theta_1 \leq \frac{\alpha_7 - \alpha_1}{6} < \frac{\pi}{6}.$$

Taking $x = \tan\theta_2$, $y = \tan\theta_1$, then

$$0 \le \frac{x-y}{1+xy} = \frac{\tan\theta_2 - \tan\theta_1}{1 + \tan\theta_2 \tan\theta_1} = \tan(\theta_2 - \theta_1) < \tan\frac{\pi}{6} = \frac{\sqrt{3}}{3}.$$

4. Divide the rectangle of 3×4 into five parts as shown in Fig. 4(Ex).1. The distance between any two points in any part is not greater than $\sqrt{5}$ by the Pythagorean theorem. This proves the conclusion.

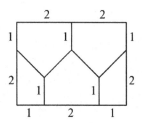

Figure 4(Ex).1

5. For each $n \in A$, $2n+2 \in B$, so, $2n+2 \le 100$, that is, $n \le 49$. Namely, $A \subset \{1, 2, \ldots, 49\}$. Divide $\{1, 2, \ldots, 49\}$ into 33 subsets as follows:

$$\{2k-1, 4k\}, k = 1, 2, \ldots, 12;$$
$$\{2k-1\}, k = 13, 14, \ldots, 25;$$
$$\{2,6\}, \{10, 22\}, \{14, 30\}, \{18, 38\};$$
$$\{2,6\}, \{34\}, \{42\}, \{46\}.$$

If $|A| > 33$, by the drawer principle, there are two numbers of A in a subset, that is, there exist n and $2n+2$ in A, and this contradicts $A \cap B = \emptyset$. Hence, $|A| \le 33$. Therefore, $|A \cup B| \le 66$.

On the other hand, if we take

$$A = \{1, 3, 5, \ldots, 49, 2, 10, 14, 18, 26, 34, 42, 46\},$$
$$B = \{2n+2 | n \in A\},$$

then A, B satisfy the properties of Problem 5, and $|A \cup B| = 66$. Summing up, we see that $|A \cup B|_{\max} = 66$.

6. Denote the language score i and maths score j of a student by the ordered pair (i, j). Then all possible scores of students form the set

$$S = \{(i, j) | 0 \le i \le 100,\ 0 \le j \le 100,\ i, j \in \mathbb{N}\}.$$

We see that S is the union of disjoint 201 subsets $A_k = \{(i, j) | (i, j) \in S,\ j = i + k\}$, $k = 0, \pm 1, \pm 2, \ldots, \pm 100$.

Suppose that there are 401 students take writing an exam. Since $A_{100} = \{(0, 100)\}$, $A_{-100} = \{(100, 0)\}$ are singletons, there are at least 399 students whose scores are in A_k ($-99 \le k \le 99$). Since $399 = 199 \times 2 + 1$, by the drawer principle, at least three scores belong to a subset, and they can be ordered by A, B, and C.

On the other hand, if the number of students is 400, and the set of their scores is

$$T = \{(i, j) | (i, j) \in S,\ \max\{i, j\} \ge 99\},$$

then there do not exist three students A, B and C, such that A is better than B and B is better than C.

Summing up, the minimum of n is 401.

7. Let $M = \{1, 2, \ldots, 50\}$.

Lemma. *The number of rows and columns that contain $i \in M$ is at least 15.*

Proof. If $i \in M$ appears at x rows, by the drawer principle, at some row there are at least $\lceil \frac{50}{x} \rceil$ of i's, so, the number of rows and columns that contain i is $S_i \ge x + \frac{50}{x} \ge 2\sqrt{50} > 14$, this proves the lemma.

We label all the rows and columns that contain $i \in M$, for $i = 1, 2, \ldots, 50$. So, we label rows and columns at least $50 \times 15 = 750$ times. But the grid has only $50 + 50 = 100$ rows and columns. By the drawer principle, there is a row or column that is labeled at least eight times. This row or column has at least eight distinct numbers. $\square$

8. **Solution 1.** Obviously, each $S(i, j) \in \mathbb{N}^+$. We shall prove that for any $n_0 \in \mathbb{N}^+$, there exists some $S(i, j) = n_0$.

Denote S_n the sum of the first n terms of $\{a_n\}$. Then

$$S_1 < S_2 < \cdots < S_{10n_0 + 10}, \tag{1}$$

$$S_1 + n_0 < S_2 + n_0 < \cdots < S_{10n_0 + 10} + n_0. \tag{2}$$

Since $S_{10n_0+10} = \sum_{k=1}^{n_0+1}(a_{10k-9} + a_{10k-8} + \cdots + a_{10k}) \le 19(n_0+1)$, and,

$$S_{10n_0+10} + n_0 \le 20n_0 + 19.$$

There are $10n_0 + 10$ distinct positive integers in (1) and $10n_0 + 10$ distinct positive integers in (2), and they are all not greater than $20n_0 + 19$. By the drawer principle, there exist two equal numbers. That is, there exist $i, j \in \mathbb{N}^+$, such that $S_j = S_i + n_0$. That is, there exist $i < j$, such that $S(i, j) = S_j - S_i = n_0$.

Therefore, the set of all $S(i, j)$ is $\mathbb{N}^+$.

Solution 2. Obviously, each $S(i, j) \in \mathbb{N}^+$. For any $n_0 \in \mathbb{N}^+$, construct $19n_0$ pairs $A_{pq} = \{2n_0p + q, (2n_0p + q) + n_0\}$, $p = 0, 1, 2, \ldots, 18, q = 1, 2, \ldots, n_0$.

Then, $\bigcup_{p=0}^{18} \bigcup_{q=1}^{n_0} A_{pq} = \{1, 2, \ldots, 38n_0\}$.

Denote S_n the sum of first n terms of $\{a_n\}$, then

$$1 \le S_1 < S_2 < \cdots < S_{20n_0} = \sum_{k=1}^{2n_0}(a_{10k-9} + a_{10k-8} + \cdots + a_{10k}) \le 38n_0.$$

By the drawer principle, there exist two numbers S_i and $S_j, 1 \le i < j \le 20n_0$ in a pair A_{pq} such that $S(i, j) = S_j - S_i = n_0$.

So, the set of all $S(i, j)$ is $\mathbb{N}^+$.

9. Denote b_k the remainder of a_k divided by $m(0 \le b_k \le m - 1)$, $k \in \mathbb{N}^+$. Consider $m^3 + 1$ triples (b_1, b_2, b_3), $(b_2, b_3, b_4), \ldots, (b_{m^3+1}, b_{m^3+2}, b_{m^3+3})$, which take at most m^3 distinct values, since $b_k \in \{0, 1, \ldots, m - 1\}$. By the drawer principle, there are two identical triples, say, $(b_i, b_{i+1}, b_{i+2}) = (b_j, b_{j+1}, b_{j+2}), (1 \le i < j \le m^3 + 1)$.

Since $a_{n+3} = a_{n+2} + 2a_{n+1} + a_n$, we have $b_k \equiv b_{k+3} - b_{k+2} - 2b_{k+1}(\bmod m)$, $k \in \mathbb{N}^+$.

So, $b_{i+2} = b_{j+2}$, $b_{i+1} = b_{j+1}$, $b_i = b_j \Rightarrow b_{i-p} = b_{j-p}$, $p = 1, 2, \ldots, i - 1$. Let $n = j - i$, and we have $b_{n+1} = b_{j-i+1} = b_1 = 1$, $b_{n+2} = b_2 = 1$, $b_{n+3} = b_3 = 3$. Thus, $b_n \equiv b_{n+3} - b_{n+2} - 2b_{n+1} \equiv 0(\bmod m)$, that is $b_n = 0$. Therefore, $m|a_n$.

10. (1) Consider an inscribed regular pentagon whose five vertices are coloured in two colours. By the drawer principle, there exist three vertices in the same colour and they are vertices of an isosceles triangle.

(2) If there are $N + 1$ points $A_1, A_2, \ldots, A_{N+1}$ on a circle in order such that the length of arc is $a > 0$ for each $i(1 \le i \le N)$, then we call such points a *group*.

Divide the semicircle into $N^2 + 1$ arcs $L_i(1 \le i \le N^2 + 1)$. If a is sufficiently small, we can make a *group* on each arc L_i. By the drawer principle, there are two points in each *group* that are in the same colour, denoted by (c_i, l_i), where c_i is the colour, and l_i the arc length between the two points. By the drawer principle, there are $i \ne j$, such that $(c_i, l_i) = (c_j, l_j)$, therefore, there are four points in the same colour, which are the vertices of an isosceles trapezoid.

Chapter 5

1. Let I be the set of all numbers of n digits without digits $0, 4, 5, \ldots, 9$. For $k = 1, 2, 3$, denote the subset of I in which no digit k appears by A_k. Obviously, the numbers satisfying the condition of the problem are $\overline{A_1} \cap \overline{A_2} \cap \overline{A_3}$. Since

$$|I| = 3^n, \ |A_1| = |A_2| = |A_3| = 2^n,$$

$$|A_1 \cap A_2| = |A_2 \cap A_3| = |A_3 \cap A_1| = 1, \ |A_1 \cap A_2 \cap A_3| = 0,$$

by the inclusion–exclusion principle, we have

$$\left| \overline{A_1} \cap \overline{A_2} \cap \overline{A_3} \right| = |I| - |A_1| - |A_2| - |A_3| + |A_1 \cap A_2|$$

$$+ |A_2 \cap A_3| + |A_3 \cap A_1| - |A_1 \cap A_2 \cap A_3|$$

$$= 3^n - 3 \cdot 2^n + 3.$$

2. Let $a_{2005} = n$. Denote $s = \{1, 2, \ldots, n\}$, $A_i = \{k \mid k \in s, \ i \mid k\}$, $i = 3, 4, 5$. Denote $\bar{A}_i$ the complement of A_i in s. Then

$$2005 = |(\bar{A}_3 \cap \bar{A}_4 \cap \bar{A}_5) \cup A_5|$$

$$= |\bar{A}_3 \cap \bar{A}_4 \cap \bar{A}_5| + |A_5|$$

$$= |s| - |A_3| - |A_4| - |A_5| + |A_3 \cap A_4| + |A_3 \cap A_5| + |A_4 \cap A_5|$$

$$- |A_3 \cap A_4 \cap A_5| + |A_5|$$

$$= n - \left\lfloor \frac{n}{3} \right\rfloor - \left\lfloor \frac{n}{4} \right\rfloor - \left\lfloor \frac{n}{5} \right\rfloor + \left\lfloor \frac{n}{12} \right\rfloor + \left\lfloor \frac{n}{15} \right\rfloor + \left\lfloor \frac{n}{20} \right\rfloor - \left\lfloor \frac{n}{60} \right\rfloor + \left\lfloor \frac{n}{5} \right\rfloor. \tag{1}$$

Since $x - 1 < \lfloor x \rfloor \le x$, by (1),

$$\begin{cases} 2005 < n - (n/3 - 1) - (n/4 - 1) + n/12 + n/15 + n/20 - (n/60 - 1), \\ 2005 > n - n/3 - n/4 + (n/12 - 1) + (n/15 - 1) + (n/20 - 1) - n/60, \end{cases}$$

which yields $3336\frac{2}{3} < n < 3346\frac{2}{3}$. Thus, $3337 \le n \le 3346$. Hence, all possible values of n are 3337, 3338, 3340, 3341, 3343, 3345, 3346.

Substitute $n = 3341$ into (1), the equality holds, and n is unique, so $a_{2005} = 3341$.

3. Since $2^{100} + 2^{100} + 2^{|C|} = 2^{|A \cup B \cup C|}$, $1 + 2^{|C| - 101} = 2^{|A \cup B \cup C| - 101}$. Hence, $|C| = 101$ and

$$|A \cup B \cup C| = 102.$$

By the inclusion–exclusion principle:

$$|A \cup B \cup C| = |A| + |B| + |C| - |A \cap B| - |B \cap C| - |C \cap A| + |A \cap B \cap C|,$$

and so,

$$|A \cap B \cap C| = 102 - (100 + 100 + 101) + |A \cap B| + |B \cap C| + |C \cap A|$$

$$= |A \cap B| + |B \cap C| + |C \cap A| - 199.$$

Meanwhile, $102 = |A \cup B \cup C| \ge |A \cup B| = |A| + |B| - |A \cap B| = 200 - |A \cap B|$. That is, $|A \cap B| \ge 98$. In the same way, $|A \cap C| \ge 99$ and $|B \cap C| \ge 99$. Thus,

$$|A \cap B \cap C| \ge 98 + 99 + 99 - 199 = 97.$$

If $A = \{1, 2, 3, \ldots, 100\}$, $B = \{3, 4, 5, \ldots, 102\}$, and $C = \{1, 2, 3, \ldots, 102\}$, then

$$|A \cap B \cap C| = |\{4, 5, \ldots, 100\}| = 97.$$

Therefore, the minimum of $|A \cap B \cap C|$ is 97.

4. Let the region of the cuboid be $\{(x, y, z) | 0 \le x \le a, 0 \le y \le b, 0 \le z \le c\}$. Parameterize the route of the particle from $(0, 0, 0)$ to (a, b, c) along the diagonal by

$$x = at, \quad y = bt, \quad z = ct, \quad t \in [0, 1].$$

When t increases from 0 to 1, if and only if at least one of x, y, and z is an integer, the particle will go into a new small cube. Let

$$A_x = \{t \mid t \in [0, 1), \ at \in \mathbb{Z}\},$$

$$A_y = \{t \mid t \in [0,1), \ bt \in \mathbb{Z}\},$$

$$A_z = \{t \mid t \in [0,1), \ ct \in \mathbb{Z}\}.$$

Then, $|A_x| = a$, $|A_y| = b$, $|A_z| = c$.

Denote $d_1 = (a,b)$, $d_2 = (b,c)$, $d_3 = (c,a)$, $d = (a,b,c)$, then,

$$A_x \cap A_y = \{t \mid t \in [0,1), \ d_1 t \in \mathbb{Z}\},$$

hence,

$$|A_x \cap A_y| = d_1.$$

In a similar way, $|A_y \cap A_z| = d_2$, $|A_z \cap A_x| = d_3$, $|A_x \cap A_y \cap A_z| = d$.
By the inclusion–exclusion principle, the number of small cubes the particle goes through is equal to

$$|A_x \cup A_y \cup A_z| = |A_x| + |A_y| + |A_z| - |A_x \cap A_y| - |A_y \cap A_z|$$

$$- |A_z \cap A_x| + |A_x \cap A_y \cap A_z|$$

$$= a + b + c - (a,b) - (b,c) - (c,a) + (a,b,c).$$

5. Denote the area of the region S_i by $|S_i|$. We see that, $|S_1 \cup S_2 \cup \cdots \cup S_{2001}| \leq 1001$, $|S_i| = 1 \ (1 \leq i \leq 2001)$.
   By the inclusion–exclusion principle,

$$1001 \geq |S_1 \cup S_2 \cup \cdots \cup S_{2001}| \geq \sum_{i=1}^{2001} |S_i|$$

$$- \sum_{1 \leq i < j \leq 2001} |S_i \cap S_j|$$

$$= 2001 - \sum_{1 \leq i < j \leq 2001} |S_i \cap S_j|$$

$$\geq 2001 - \frac{2001 \times 2000}{2} \cdot S,$$

where $S = \max\limits_{1 \leq i < j \leq 2001} |S_i \cap S_j|$.

That is, $S \geq \frac{1}{2001}$.

Remark. This is the area overlapping form of the inclusion–exclusion principle.

6. Let the five posts equipped with new weapons be at the positions $a_1, a_2, a_3, a_4, a_5 \in \{1, 2, \ldots, 20\}$ in increasing order. Denote

$$x_1 = a_1, x_2 = a_2 - a_1, x_3 = a_3 - a_2, x_4 = a_4 - a_3, x_5 = a_5 - a_4, x_6 = 20 - a_5.$$

Then,

$$x_1 + x_2 + x_3 + x_4 + x_5 + x_6 = 20, \ 2 \le x_k \le 5 \, (1 \le k \le 5), \ 1 \le x_6 \le 4.$$

By the substitution $y_k = x_k - 1 \, (1 \le k \le 5)$, $y_6 = x_6$, we have

$$y_1 + y_2 + y_3 + y_4 + y_5 + y_6 = 15, \ 1 \le y_k \le 4 \, (1 \le k \le 6).$$

Let I be the set of positive integer solutions of the Diophantine equation

$$y_1 + y_2 + y_3 + y_4 + y_5 + y_6 = 15.$$

Let A_k be the subset of I such that $y_k > 4$. Then

$$\left| \bigcap_{k=1}^{6} \overline{A_k} \right| = |I| - \left| \bigcup_{k=1}^{6} A_k \right| = |I| - \sum_{k=1}^{6} |A_k| + \sum_{1 \le j < k \le 6} |A_j \cap A_k|,$$

where $A_i \cap A_j \cap A_k = \varnothing$, since there is no solution in I satisfying $y_i, y_j, y_k > 4$. By the correspondence principle (cf. Chapter 12), we see that $|I| = \binom{14}{5}$, $|A_k| = \binom{10}{5}$, and $|A_j \cap A_k| = \binom{6}{5}$. Therefore,

$$\left| \bigcap_{k=1}^{6} \overline{A_k} \right| = \binom{14}{5} - 6 \binom{10}{5} + \binom{6}{2}\binom{6}{5} = 2002 - 1512 + 90 = 580.$$

Since the number of the permutations of 5 weapons for 5 posts is $5! = 120$, the answer of Problem 6 is $580 \times 120 = 69,600$.

7. Let P_0 be the set of all ordered groups $(A_1, A_2, \ldots, A_m)$ satisfying $\bigcup_{i=1}^{m} A_i = \{1, 2, \ldots, n\}$. For $1 \le i \le m$, let $P_i = \{(A_1, A_2, \ldots, A_m) \in P_0 | A_i = \varnothing\}$.

For each $x \in \{1, 2, \ldots, n\}$, there are 2^m cases of whether or not x belongs to each of $A_1, A_2, \ldots, A_m$, and where the number of cases such that $x \in \bigcup_{i=1}^{m} A_i$ is $2^m - 1$. By the multiplication principle, we have $|P_0| = (2^m - 1)^n$.

For $1 \le i_1 < i_2 < \cdots < i_s \le m$, in a similar way,

$$|P_{i_1} \cap P_{i_2} \cap \cdots \cap P_{i_s}| = (2^{m-s} - 1)^n.$$

By the inclusion–exclusion principle,

$$f(m, n) = \left| \bigcap_{k=1}^{m} \overline{P_k} \right| = |P_0| - \sum_{i=1}^{m} |P_i| + \sum_{1 \le i < j \le m} |P_i \cap P_j| - \cdots + (-1)^m$$
$$\times |P_1 \cap P_2 \cap \cdots \cap P_m|$$

$$= (2^m - 1)^n - \binom{m}{1}(2^{m-1} - 1)^n + \binom{m}{2}(2^{m-2} - 1)^n - \cdots$$

$$+ (-1)^m \binom{m}{m}(2^{m-m} - 1)^n$$

$$= \sum_{k=0}^{m-1} (-1)^k \binom{m}{k}(2^{m-k} - 1)^n.$$

Remark. Now, here is an interesting result: $f(m, n) = f(n, m)$. In fact, if we consider all $m \times n$ matrices with only elements 0 and 1, we can prove by the correspondence principle (cf. Chapter 12) that the number of matrices with 1 appearing in each row and each column is equal to both $f(m, n)$ and $f(n, m)$.

8. Consider the universal set $I_n = \{1, 2, \ldots, n\}$. Let

$$A_n = S(a) \cap I_n, B_n = S(b) \cap I_n, C_n = S(c) \cap I_n.$$

By the inclusion–exclusion principle, we have

$$|A_n \cup B_n \cup C_n| = |A_n| + |B_n| + |C_n| - |A_n \cap B_n|$$
$$- |B_n \cap C_n| - |C_n \cap A_n| + |A_n \cap B_n \cap C_n|.$$

Since $|A_n \cup B_n \cup C_n| \le n$ and $|A_n \cap B_n \cap C_n| \ge 0$, and note that

$$|A_n| \ge \left\lfloor \frac{n}{a} \right\rfloor > \frac{n}{a} - 1, \; |B_n| \ge \left\lfloor \frac{n}{b} \right\rfloor > \frac{n}{b} - 1, \; |C_n| \ge \left\lfloor \frac{n}{c} \right\rfloor > \frac{n}{c} - 1,$$

therefore,

$$n > \left(\frac{n}{a} + \frac{n}{b} + \frac{n}{c} - 3\right) - |A_n \cap B_n| - |B_n \cap C_n| - |C_n \cap A_n|.$$

Consequently,

$$|A_n \cap B_n| + |B_n \cap C_n| + |C_n \cap A_n| > \left(\frac{1}{a} + \frac{1}{b} + \frac{1}{c} - 1\right) n - 3.$$

Since $\frac{1}{a} + \frac{1}{b} + \frac{1}{c} > 1$, the right-hand side of the above inequality tends to ∞ as $n \to \infty$. By definition, for any $n \in \mathbb{N}^+$, $A_n \subseteq S(a)$, $B_n \subseteq S(b)$, $C_n \subseteq S(c)$. Therefore, there are two sets of $S(a), S(b), S(c)$ whose intersection is an infinite set.

Chapter 6

1. Suppose that player A wins the most matches. Since A does not win all matches, there is a player C who beats A. Then there must be a player B among those beaten by A who beats C (otherwise, C wins more than A). This completes the proof.
2. There is a maximal distance between every two of the 100 points (Property 1). Say, the distance between A and B is the largest (if there are several, just take any one). Draw a circle O with diameter $AB(\leq 1)$.

    Take any point C in the other 98 points. Since $AB \geq AC$, $AB \geq BC$, and $\triangle ABC$ is obtuse, as in Fig. 6(Ex).1, we see that $\angle C$ is obtuse. Thus, C is inside the circle O. That is, the circle O covers all the 100 points.

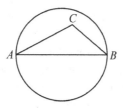

Figure 6(Ex).1

3. Suppose that point A is incident to the most n (≤ 7) edges.

    If there is no triangle formed by 3 of 17 edges, then there are no edges between the n points adjacent to A. And each point of the other $(7-n)$ points is incident to at most n edges. Thus, the total number of edges does not exceed

$$n + (7 - n)n = -n^2 + 8n = -(n - 4)^2 + 16 \leq 16.$$

A contradiction.

Remark. The result of Problem 3 can be enhanced to *have at least 4 triangles*, a hard problem left to the interested reader.

4. Suppose that A knows the most n people B_1, B_2, ..., B_n in the party.

   Since A is known by both B_i and B_j among B_1, B_2, ..., B_n, the numbers of people known by B_i and by B_j are different. Thus, the numbers of people known by B_1, B_2, ..., B_n are distinct, and they are exactly $1, 2, \ldots$, and n since they are $\leq n$.

   Given that someone knows at least 2012 people, it means $n \geq 2012$. So, someone among B_1, B_2, ..., B_n knows 2012 people.

5. Among n distances from point P to the sides of convex n-gon, there is a shortest one. Suppose it is the segment PQ and point Q is on line AB. We show that Q is on the side AB but $Q \neq A$ or B.

   Suppose, on the contrary, then PQ meets some side other than side AB.

   Say, PQ meets side MN at point R and PR is not perpendicular to MN, then the distance d from P to line MN satisfies $d < PR \leq PQ$, contradicting the choice of PQ as the shortest distance.

6. Let l be the shortest route through all cities, and it goes from city A to city B. Then the numbers on the boards of city A and city B are N. For any city C, it must be on the route l. So, the distance along l from C to A or to B is not greater than $N/2$, say, it is the city A and it goes through all the cities. Then, along the route l first from C to A then from A to B, distance is not greater than $3N/2$. Thus, the number on the board of any city is not greater than $3N/2$ and not less than N. This completes the proof.

7. Suppose, on the contrary, there exist regular n-gons with integral vertices. Then, their areas are positive integers or half integers. Suppose the regular n-gon $A_1 A_2 \ldots A_n$ with side length a and circumradius R has the smallest area.

   Since $n \geq 7$, $\frac{2\pi}{n} < \frac{\pi}{3}$. So $a < R$.

   Translate the vector $\overrightarrow{A_i A_{i+1}}$ to the vector $\overrightarrow{OB_i}(i = 1, 2, \ldots, n)$, where O is the origin. Then B_1, B_2, ..., B_n are all integral points, and the angles of $\angle B_i O B_{i+1}$ are the same ($A_{n+1} = A_1, B_{n+1} = B_1$). Thus,

$B_1 B_2 \ldots B_n$ is also an integral regular n-gon, with circumradius $a < R$. So $B_1 B_2 \ldots B_n$ has smaller area than $A_1 A_2 \ldots A_n$. A contradiction.

Remark. This problem is solved by using the idea of infinite descending, that is, under the premise of the existence of an integral regular n-gon, we can always construct another integral regular n-gon smaller than the current one, but the area of integral n-gons cannot be infinitesimally small, thus leading to the contradiction.

8. Suppose, on the contrary, under the given condition, there exist 2007 integers x_1, $x_2, \ldots, x_{2007}$ on the circle anticlockwise, such that $|x_1| + |x_2| + \cdots + |x_{2007}|$ attains its minimal.

   Since in any adjacent five numbers there are three numbers whose sum is twice the sum of the other two numbers, $x_i + x_{i+1} + x_{i+2} + x_{i+3} + x_{i+4} \equiv 0 \,(\mathrm{mod}\,3)$, $i = 1, 2, \ldots, 2007$. Here, if $j \equiv k \,(\mathrm{mod}\,2007)$, then $x_j = x_k$. Thus, $x_i \equiv x_{i+5} \,(\mathrm{mod}\,3)$.

   Since 5 and 2007 are coprime, $5m - 2007n = 1$ has integer solutions. Hence, $x_i \equiv x_1 \,(\mathrm{mod}\,3)$, for $i = 1, 2, \ldots, 2007$. It follows that

   $$x_i + x_{i+1} + x_{i+2} + x_{i+3} + x_{i+4} \equiv 5x_1 \equiv 0 \,(\mathrm{mod}\,3),$$

   yielding, $x_1 \equiv 0 \,(\mathrm{mod}\,3)$. Therefore, $x_1, x_2, \ldots, x_{2007}$ are all multiples of 3.

   Let $y_i = \frac{x_i}{3}, i = 1, 2, \ldots, 2007$, then $y_1, y_2, \ldots, y_{2007}$ are also the 2007 integers satisfying the condition of the problem, while $|y_1| + |y_2| + \cdots + |y_{2007}| = (|x_1| + |x_2| + \cdots + |x_{2007}|)/3$, a contradiction.

9. Suppose, on the contrary, $m > \sqrt{n/3}$ is the least positive integer such that $m^2 + m + n$ is a composite number.

   If $m \le n - 2$, then $m^2 + m + n \le (n-2)(n-1) + n < n^2$.

   Let p be the least prime factor of $m^2 + m + n$, then $p \le \sqrt{m^2 + m + n} < n$.

   (I) If $m \ge p$, then $(m - p)^2 + (m - p) + n$ is a positive multiple of p and larger than P. A contradiction.

   (II) If $m \le p - 1$, then

   $$p < (p - 1 - m)^2 + (p - 1 - m) + n = p^2 - 2p(m+1) + m^2 + m + n$$

   is a composite number, since it can be divided by p. Under the premise that m is the minimum, we see that $p - 1 - m \ge m$. Hence,

   $$2m + 1 \le p \le \sqrt{m^2 + m + n}.$$

That is, $3m^2 + 3m + 1 \le n$. On the other hand, since $m > \sqrt{\frac{n}{3}}$, we have $3m^2 + 3m + 1 > 3m^2 > n$. A contradiction.

Remark. Under the assumption of the contrary, there exist the smallest m and the smallest prime factor p of $m^2 + m + n$, and we derive a smaller m by adjustment and which leads to the contradiction.

Chapter 7

1. Congruence modulo 2 has some interesting properties:

$$-a \equiv a \pmod 2 \quad \text{and} \quad |a| \equiv a \pmod 2.$$

By these properties, we see that

$$b_1 + b_2 + \cdots + b_{27}$$

$$\equiv |a_1 - a_2 + a_3| + |a_4 - a_5 + a_6| + \cdots + |a_{79} - a_{80} + a_{81}|$$

$$\equiv a_1 + a_2 + a_3 + a_4 + a_5 + a_6 + \cdots + a_{79} + a_{80} + a_{81}$$

$$\equiv a_1 + a_2 + \cdots + a_{81} \equiv 1 + 2 + \cdots + 81 \equiv 41 \times 81 \equiv 1 \pmod 2.$$

Thus,

$$|b_1 - b_2 + b_3| + |b_4 - b_5 + b_6| + \cdots + |b_25 - b_26 + b_{27}|$$

$$\equiv b_1 + b_2 + b_3 + b_4 + b_5 + b_6 + \cdots + b_{25} + b_{26} + b_{27} \equiv 1 \pmod 2.$$

Continue this process, finally, $x = 1 \pmod 2$, namely, x is odd.

2. First we show that n is even.

Suppose, on the contrary that n is odd, then $a_1, a_2, \ldots, a_n$ are all odd since $a_1 a_2 \ldots a_n = n$ is odd. Then,

$$a_1 + a_2 + \cdots + a_n \equiv n \equiv 1 \pmod 2.$$

Contradicting $a_1 + a_2 + \cdots + a_n = 0$.

Now we show that $4|n$.

Suppose, on the contrary, then exactly one of $a_1, a_2, \ldots, a_n$ is even. Thus,

$$a_1 + a_2 + \cdots + a_n \equiv n - 1 \equiv 1 \pmod 2,$$

contradicting to $a_1 + a_2 + \cdots + a_n = 0$. So, $4|n$.

3. (a) The sum of six odd numbers in (1) is even.

   (b) Since the product of the six terms in (1) is

$$-(rstuvwxyz)^2 = -1,$$

at least one term is -1. So, the maximum of (1) is at most 4. If we let u, x, and y be -1, and all the others be 1, then the value of (1) is 4. Therefore, the maximum of (1) is 4.

4. The answer is negative. To show this, let Array C be obtained by removing all the absolute value signs in Array B:

C_{11}	C_{12}	C_{13}
C_{21}	C_{22}	C_{23}
C_{31}	C_{32}	C_{33}

C

Then,

$$C_{11} = (a_{11} + a_{12} + a_{13}) - (a_{11} + a_{21} + a_{31}),$$
$$C_{12} = (a_{11} + a_{12} + a_{13}) - (a_{12} + a_{22} + a_{32}),$$
$$\cdots,$$
$$C_{33} = (a_{31} + a_{32} + a_{33}) - (a_{13} + a_{23} + a_{33}).$$

It is easy to see, $C_{11} + C_{12} + \cdots + C_{33} = 0$, so there are an even number of odd numbers in Array C. Since $b_{ij} = |C_{ij}|$ and C_{ij} have the same parity, there are an even number of odd numbers in Array B. However, there are an odd number of odd numbers in $1, 2, \ldots, 9$. Thus, the answer is negative.

5. Since all positive factors of an odd number are odd, $d_1^2 + d_2^2 + d_3^2 + d_4^2 = n$ is even. So, n must be even, and $d_1 = 1, d_2 = 2$.

   If $4|n$, then one of d_3 and d_4 is 4 and the other is odd. We have

$$d_1^2 + d_2^2 + d_3^2 + d_4^2 \equiv 2 \pmod 4,$$

contradicting $4|n$.

Therefore, $n = 2m$, where m is odd.

Obviously, d_3 is the least odd prime factor of m, and d_4 is even. So $d_4 = 2d_3$.

Thus, $n = 1^2 + 2^2 + d_3^2 + 4d_3^2 = 5(1 + d_3^2)$. That is, $5|n$. Hence, d_3 cannot be 3. Let $d_3 = 5$. We check that $n = 1^2 + 2^2 + d_3^2 + 4d_3^2 = 5(1 + 5^2) = 130$ meets the requirement.

6. We discuss in two cases:

   (I) If $n(> 2)$ is odd, let $n = 2k + 1(k \in \mathbb{N}^+)$. Since the absolute value of the difference of $k + 1$ and its neighbouring number is no more than k; we have $f(s) \leq k$; on the other hand, $f(s) = k = \frac{n-1}{2}$ for the permutation $(k + 1, 1, k + 2, 2, \ldots, 2k, k, 2k + 1)$.

   (II) If n is even, let $n = 2k(k \in \mathbb{N}^+)$. Similarly, the absolute value of the difference of $k + 1$ and its neighbouring number is no more than k. And $f(s) \leq k$. And $f(s) = k = \frac{n}{2}$ for the permutation $(k + 1, 1, k + 2, 2, \ldots, 2k, k)$, $f(s) = k = \frac{n}{2}$.

   Summing up, the maximum of $f(s)$ is $\lfloor \frac{n}{2} \rfloor$.

7. First, we show that there are infinitely many odd numbers in $\{a_n\}$.

   Suppose, on the contrary, there is a last odd term a_m in $\{a_n\}$. Let $a_{m+1} = 2^p \cdot q$ with $p \in \mathbb{N}^+$ and q odd.

$$a_{m+2} = \left\lfloor \frac{3a_{m+1}}{2} \right\rfloor = 3 \cdot 2^{p-1} \cdot q,$$

$$a_{m+3} = \left\lfloor \frac{3a_{m+2}}{2} \right\rfloor = 3^2 \cdot 2^{p-2} \cdot q,$$

$$\cdots\cdots$$

Thus, $a_{m+p+1} = \left\lfloor \frac{3a_{m+p}}{2} \right\rfloor = 3^p \cdot q$ is odd. A contradiction.

Then we show that there are infinitely many even numbers in $\{a_n\}$.

Suppose, on the contrary, that a_l is the last even number in $\{a_n\}$. Let $a_{l+1} = 2^p \cdot q + 1$ with $p \in \mathbb{N}^+$ and q odd. Then,

$$a_{l+2} = \left\lfloor \frac{3a_{l+1}}{2} \right\rfloor = \left\lfloor 3 \cdot 2^{p-1} \cdot q + \frac{3}{2} \right\rfloor = 3 \cdot 2^{p-1} \cdot q + 1,$$

$$a_{l+3} = \left\lfloor \frac{3a_{l+2}}{2} \right\rfloor = 3^2 \cdot 2^{p-2} \cdot q + 1,$$

$$\cdots\cdots$$

Thus, $a_{l+p+1} = \left\lceil \frac{3a_{l+p}}{2} \right\rceil = 3^p \cdot q + 1$ is even. A contradiction.

8. Let $n = p_1^{\alpha_1} p_2^{\alpha_2} \cdots p_k^{\alpha_k}$ be the prime factorization of n, then the sum of all positive factors can be expressed by

$$(1 + p_1 + \cdots + p_1^{\alpha_1})(1 + p_2 + \cdots + p_2^{\alpha_2}) \cdots (1 + p_k + \cdots + p_k^{\alpha_k}).$$

If it is a power of 2, then each factor $f_i = 1 + p_i + \cdots + p_i^{\alpha_i}$ $(i = 1, 2, \ldots, k)$ is a power of 2, which implies that all p_i and α_i are odd.

If there is an $\alpha_i > 1$, then

$$f_i = (1 + p_i)(1 + p_i^2 + p_i^4 + \cdots + p_i^{\alpha_i - 1}).$$

Here, the second factor is even, hence $(\alpha_i - 1)/2$ is odd. Consequently, $1 + p_i^2$ is also a factor of f_i. Since $1 + p_i$ and $1 + p_i^2$ are all powers of 2, we see that $(1 + p_i) | (1 + p_i^2)$. On the other hand,

$$1 + p_i^2 = (1 + p_i)(p_i - 1) + 2.$$

That is, $1 + p_i^2 \equiv 2 \pmod{1 + p_i}$, a contradiction. Hence, $\alpha_i = 1 (i = 1, 2, \ldots, k)$.

Therefore, the number of positive factors of n is a power of 2.

9. Denote $a_1, a_2, \ldots, a_{2n+1}$ as the number of balls in the $(2n + 1)$ bags. Without loss of generality, suppose that $a_1 \leq a_2 \leq \cdots \leq a_{2n+1}$. The problem is equivalent to the following problem: take out any number from $(2n+1)$ nonnegative integers, the rest $2n$ numbers can be divided into two groups of n numbers with equal sum. Show these $2n+1$ numbers are all equal.

Let $A = a_1 + a_2 + \cdots + a_{2n+1}$, then for each i $(1 \leq i \leq 2n+1)$, $A - a_i$ is even (otherwise, the remaining numbers cannot be divided into two groups with equal sum). Thus, a_i and A have the same parity, and so do $a_1, a_2, \ldots, a_{2n+1}$.

It is easy to see that $(2n + 1)$ numbers

$$0, a_2 - a_1, a_3 - a_1, \ldots, a_{2n+1} - a_1$$

also satisfy the property. Since $a_i - a_1$ $(i = 2, 3, \ldots, 2n + 1)$ is even, $(2n + 1)$ numbers

$$0, \frac{a_2 - a_1}{2}, \frac{a_3 - a_1}{2}, \ldots, \frac{a_{2n+1} - a_1}{2}$$

also satisfy the property and are all even.

This process cannot go on forever, unless

$$a_1 = a_2 = \cdots = a_{2n+1}.$$

That is, all bags contain the same number of balls.

Chapter 8

1. Let $S_{\triangle PP_2P_3} = S_1, S_{\triangle PP_3P_1} = S_2, S_{\triangle PP_1P_2} = S_3$.

   It is easy to see that $\dfrac{P_1P}{PQ_1} = \dfrac{S_{\triangle P_1P_2P}}{S_{\triangle PP_2Q_1}} = \dfrac{S_{\triangle P_1P_3P}}{S_{\triangle PP_3Q_1}}$, so

   $$\frac{P_1P}{PQ_1} = \frac{S_{\triangle P_1P_2P} + S_{\triangle P_1P_3P}}{S_{\triangle PP_2Q_1} + S_{\triangle PP_3Q_1}} = \frac{S_2 + S_3}{S_1}.$$

   Similarly, we have

   $$\frac{P_2P}{PQ_2} = \frac{S_3 + S_1}{S_2}, \quad \frac{P_3P}{PQ_3} = \frac{S_1 + S_2}{S_3}.$$

   By symmetry, we can suppose that, say, $S_1 \geq S_2 \geq S_3$, and then we have $\frac{P_1P}{PQ_1} \leq 2$, $\frac{P_3P}{PQ_3} \geq 2$.

2. Let the line l bisect the perimeter of $\triangle ABC$.

   If l passes through a vertex, say A, then it is not difficult to see that $AB = AC$, thus, l passes through the incentre I.

   If l passes two sides of $\triangle ABC$, let l meet AB and AC at E and D, respectively, as shown in Fig. 8(Ex).1. We will show that $S_{\triangle IDE} = 0$.

   Let r be the inradius of $\triangle ABC$, and p be the semi-perimeter. Then

   $$AE + AD = p,$$

   $$S_{\triangle AIE} + S_{\triangle AID} = \tfrac{1}{2}r \cdot AE + \tfrac{1}{2}r \cdot AD = \tfrac{1}{2}rp = \tfrac{1}{2}S_{\triangle ABC}.$$

   While $S_{\triangle AED} = \tfrac{1}{2}S_{\triangle ABC}$, hence $S_{\triangle IDE} = 0$.

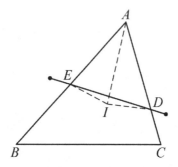

Figure 8(Ex).1

3. Let $\frac{AM}{AC} = \frac{CN}{CD} = r(0 < r < 1)$, see Fig. 8(Ex).2.

   Suppose without loss of generality that $S_{\triangle ABC} = 1$, then by the given conditions, $S_{\triangle ABD} = 3$, $S_{\triangle BCD} = 4$, $S_{\triangle ACD} = 3 + 4 - 1 = 6$.

   Consequently, we have

$$S_{\triangle BMC} = \frac{S_{\triangle BMC}}{S_{\triangle ABC}} = \frac{MC}{AC} = 1 - r,$$

$$S_{\triangle MNC} = \frac{6S_{\triangle MNC}}{S_{\triangle ACD}} = 6 \cdot \frac{MC}{AC} \cdot \frac{CN}{CD} = 6r(1 - r),$$

$$S_{\triangle NBC} = \frac{4S_{\triangle NBC}}{S_{\triangle BCD}} = \frac{4CN}{CD} = 4r.$$

   Thus, $4r = S_{\triangle NBC} = S_{\triangle MNC} + S_{\triangle BMC} = 6r(1 - r) + 1 - r$. So, $6r^2 - r - 1 = 0$, and $r = \frac{1}{2}$, since $0 < r < 1$. That is, M and N are the midpoints of AC and CD, respectively.

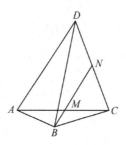

Figure 8(Ex).2

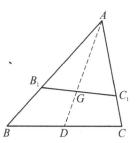

Figure 8(Ex).3

4. Make median AD on side BC as shown in Fig. 8(Ex).3. Let the area of $\triangle ABC$ be 1 and the area of $\triangle AB_1C_1$ be S. We calculate S in two ways. On the one hand,

$$S = \frac{S}{1} = \frac{AB_1 \cdot AC_1}{AB \cdot AC} = \lambda\mu. \tag{1}$$

On the other hand, $S = S_{\triangle AB_1G} + S_{\triangle AGC_1}$.
Similar to (1), $S_{\triangle AB_1G} = \frac{\lambda}{3}, S_{\triangle AGC_1} = \frac{\mu}{3}$.
So, $S = \frac{\lambda+\mu}{3}$.
Summing up, we have $\lambda\mu = \frac{\lambda+\mu}{3}$.
Multiply both sides by $\frac{3}{\lambda\mu}$, and we obtain the desired result.

5. Let QT and RU meet at point X. Draw segments XA, XB, XC, XD, XE, XF, XP, XS as shown in Fig. 8(Ex).4.
   We know that

$$S_{ABQTF} = \frac{1}{2}S_{ABCDEF} = S_{ABCRU}.$$

Subtract S_{ABQXU} from both sides to yield $S_{XTFU} = S_{XQCR}$.

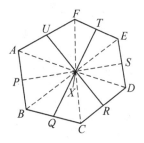

Figure 8(Ex).4

Further, we have

$$S_{XEFA} = S_{\triangle XEF} + S_{\triangle XFA} = 2S_{\triangle XTF} + 2S_{\triangle XFU} = 2S_{XTFU},$$

$$S_{XBCD} = S_{\triangle XBC} + S_{\triangle XCD} = 2S_{\triangle XQC} + 2S_{\triangle XCR} = 2S_{XQCR}.$$

Thus, $S_{XEFA} = S_{XBCD}$. Combine $S_{\triangle XAP} = S_{\triangle XBP}$, $S_{\triangle XES} = S_{\triangle XDS}$ to deduce that, polyline PXS bisects the area of the convex hexagon $ABCDEF$. Since PS bisects the area of $ABCDEF$ as well, PS passes through point X.

Thus, three lines PS, QT, and RU are concurrent.

6. Let P be the intersection point of AC and BD.

Let the included angle of AC and BD be α, and the included angle of $A'C'$ and $B'D'$ be β. By symmetry, the diagonals $A'C'$ and $B'D'$ of quadrilateral $A'B'C'D'$ meet at point P, and $A'C' = AC$, $B'D' = BD$, $\beta = 3\alpha$ or $\pi - 3\alpha$ or $3\alpha - \pi$. In either case, we have

$$\sin\beta = |\sin 3\alpha| = \sin\alpha \cdot |3 - 4\sin^2\alpha|.$$

By the area formula for quadrilaterals,

$$S' = \frac{1}{2}A'C' \cdot B'D' \cdot \sin\beta = \frac{1}{2}AC \cdot BD \cdot \sin\alpha \cdot \left|3 - 4\sin^2\alpha\right|,$$

and this implies, $\frac{S'}{S} = |3 - 4\sin^2\alpha| < 3$.

7. Join PQ, PN, and NQ as shown in Fig. 8(Ex).5.

Denote $\angle PAN = \alpha$ and $\angle QAN = \beta$. Since M is the midpoint of BC, $S_{\triangle ABC} = 2S_{\triangle ABM} = 2S_{\triangle AMC}$. That is,

$$\frac{1}{2}AB \cdot AC \cdot \sin(\alpha + \beta) = AB \cdot AM \cdot \sin\alpha = AM \cdot AC \cdot \sin\beta.$$

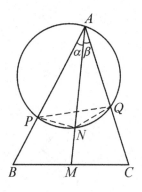

Figure 8(Ex).5

Therefore,

$$\frac{\sin \alpha}{\sin(\alpha + \beta)} = \frac{AC}{2AM}, \frac{\sin \beta}{\sin(\alpha + \beta)} = \frac{AB}{2AM}. \tag{8.1}$$

By the law of sine, we have

$$\frac{NP}{\sin \alpha} = \frac{NQ}{\sin \beta} = \frac{PQ}{\sin(\alpha + \beta)}. \tag{8.2}$$

Since A, P, N, Q are concyclic, by Ptolemy's theorem,

$$AN \cdot PQ = AP \cdot NQ + AQ \cdot NP.$$

By (8.1) and (8.2),

$$\begin{aligned}
AN &= AP \cdot \frac{NQ}{PQ} + AQ \cdot \frac{NP}{PQ} = \frac{AP \cdot \sin \beta}{\sin(\alpha + \beta)} + \frac{AQ \cdot \sin \alpha}{\sin(\alpha + \beta)} \\
&= \frac{AP \cdot AB}{2AM} + \frac{AQ \cdot AC}{2AM}.
\end{aligned}$$

Thus,

$$AP \cdot AB + AQ \cdot AC = 2AN \cdot AM.$$

8. By the common edge theorem, we have $\frac{AB'}{B'C} = \frac{S_{\triangle ABJ}}{S_{\triangle CBJ}}$ and a few other similar relations on the segment length ratios and triangle area ratios. It suffices to show that

$$\frac{S_{\triangle ABJ}}{S_{\triangle CBJ}} \cdot \frac{S_{\triangle CDG}}{S_{\triangle EDG}} \cdot \frac{S_{\triangle EAI}}{S_{\triangle BAI}} \cdot \frac{S_{\triangle BCF}}{S_{\triangle DCF}} \cdot \frac{S_{\triangle DEH}}{S_{\triangle AEH}} = 1.$$

By

$$\frac{S_{\triangle ABJ}}{S_{\triangle BAI}} = \frac{S_{\triangle ABJ}}{S_{\triangle ABD}} \cdot \frac{S_{\triangle ABD}}{S_{\triangle BAI}} = \frac{AJ}{AD} \cdot \frac{BD}{BI},$$

and some other relations after eliminating the same terms on both sides of the fraction, we need only to show that

$$\frac{AJ}{BI} \cdot \frac{BF}{CJ} \cdot \frac{CG}{DF} \cdot \frac{DH}{EG} \cdot \frac{EI}{AH} = 1,$$

or

$$\frac{AJ}{CJ} \cdot \frac{BF}{DF} \cdot \frac{CG}{EG} \cdot \frac{DH}{AH} \cdot \frac{EI}{BI} = 1.$$

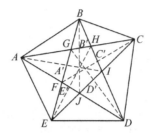

Figure 8(Ex).6

By using the law of sine, we need only to show that

$$\frac{\sin\angle ECA}{\sin\angle CAD} \cdot \frac{\sin\angle ADB}{\sin\angle DBE} \cdot \frac{\sin\angle BEC}{\sin\angle ECA} \cdot \frac{\sin\angle CAD}{\sin\angle ADB} \cdot \frac{\sin\angle DBE}{\sin\angle BEC} = 1,$$

which holds obviously (see Fig. 8(Ex).6).

Chapter 9

1. Since each transformation changes the signs of four numbers in the array, $(-1)^4 = 1$. So, the sign of the product of the four numbers in the changed row (or column) will not change under the transformation.

   At the beginning, the product of the 16 numbers in the array is negative and it cannot be changed. So, it is impossible to make all the numbers positive.

2. Consider the product of these six numbers

$$(a_1 a_5 a_9)(a_2 a_6 a_7)(a_3 a_4 a_8)(-a_3 a_5 a_7)(-a_1 a_6 a_8)(-a_2 a_4 a_9)$$
$$= -(a_1 a_2 a_3 a_4 a_5 a_6 a_7 a_8 a_9)^2 < 0.$$

   Therefore, at least one of the six numbers is negative.

3. Let $A = \{a_1, a_2, \ldots, a_t\}$ have the least number of elements. Since the number of binary subsets of A is $\binom{t}{2}$, there are at most $t + \binom{t}{2} = \frac{t(t+1)}{2}$ distinct positive integers obtained by taking the elements of A and the sums of two different elements of A.

   So, $\frac{t(t+1)}{2} \geq |S| = 21 \Rightarrow t \geq 6$.

   If $t = 6$, then each element in S has the unique form of $a_i (i = 1, 2, \ldots, 6)$ or $a_i + a_j (1 \leq i < j \leq 6)$.

   Denote $B = \{a_i + a_j | 1 \leq i < j \leq 6, i, j \in \mathbb{N}^+\}$.

   On the one hand, $\sum_{x \in S} x = 1 + 2 + \cdots + 21 = 231$.

On the other hand,

$$\sum_{x \in S} x = \sum_{x \in A} x + \sum_{x \in B} x = \sum_{i=1}^{6} a_i + \sum_{1 \leq i < j \leq 6} (a_i + a_j)$$

$$= \sum_{i=1}^{6} a_i + 5 \sum_{i=1}^{6} a_i = 6 \sum_{i=1}^{6} a_i,$$

which is even, a contradiction. So, $t \geq 7$.

If $t = 7$, we check that the set $A = \{1, 2, 3, 6, 10, 14, 18\}$ satisfies the condition.

Summing up, the least number of elements of set A is 7.

4. Let $S = x_1 + x_2 + x_3 + x_4 + x_5$. We know that x_1, x_2, x_3, x_4, x_5 all satisfy the equation

$$x^2 - Sx - 1 = 0.$$

Thus, x_1, x_2, x_3, x_4, x_5 take value in $\frac{S \pm \sqrt{S^2 + 4}}{2}$, and

$$x_1 + x_2 + x_3 + x_4 + x_5 = \frac{5S + \varepsilon\sqrt{S^2 + 4}}{2},$$

where $\varepsilon \in \{\pm 1, \pm 3, \pm 5\}$. So, $S = \frac{5S + \varepsilon\sqrt{S^2 + 4}}{2}$, that is $3S = -\varepsilon\sqrt{S^2 + 4}$.

Further, by $|3S| = |-\varepsilon\sqrt{S^2 + 4}| > |\varepsilon S|$, we see that $\varepsilon = \pm 1$. So, $S = \pm\frac{\sqrt{2}}{2}$.

If $S = \frac{\sqrt{2}}{2}$, solving $x_1^2 - \frac{\sqrt{2}}{2}x_1 - 1 = 0$, we have $x_1 = \sqrt{2}$ or $x_1 = -\frac{\sqrt{2}}{2}$; if $S = -\frac{\sqrt{2}}{2}$, we have $x_1 = -\sqrt{2}$ or $x_1 = \frac{\sqrt{2}}{2}$.

Therefore, all possible values of x_1 are $\pm\sqrt{2}$ and $\pm\frac{\sqrt{2}}{2}$.

Remark. The five given equations have obvious cyclic characteristics, which prompt us to consider the problem as a whole. So, introduce the quantity $S = x_1 + x_2 + x_3 + x_4 + x_5$, and transform the original system of equations with five arguments into a simple problem: how many times can x_1, x_2, x_3, x_4, x_5 take the two roots of $x^2 - Sx - 1 = 0$?

5. The answer is negative. Suppose it is possible. Since a quadratic trinomial has at most two real roots and the $2n$ real numbers a_i, b_i are distinct, each quadratic trinomial $x^2 - a_i x + b_i (1 \leq i \leq n)$ has two distinct real roots u_i and v_i, and $(a_1, a_2, \ldots, a_n, b_1, b_2, \ldots, b_n)$ is a permutation of $(u_1, u_2, \ldots, u_n, v_1, v_2, \ldots, v_n)$.

For $1 \leq i \leq n$, by Viète's theorem,

$$u_i + v_i = a_i, u_i v_i = b_i.$$

So,

$$\sum_{i=1}^{n} a_i = \sum_{i=1}^{n} (u_i + v_i) = \sum_{i=1}^{n} a_i + \sum_{i=1}^{n} b_i.$$

That is, $\sum_{i=1}^{n} b_i = 0$.

On the other hand, we have

$$u_i^2 + v_i^2 = (u_i + v_i)^2 - 2u_i v_i = a_i^2 - 2b_i.$$

Consequently,

$$\sum_{i=1}^{n} (a_i^2 + b_i^2) = \sum_{i=1}^{n} (u_i^2 + v_i^2) = \sum_{i=1}^{n} (a_i^2 - 2b_i) = \sum_{i=1}^{n} a_i^2.$$

This implies $\sum_{i=1}^{n} b_i^2 = 0$. That is, all b_i's are zero. A contradiction.

6. The answer is positive. Consider the sum P_k of the products of all two neighbouring numbers at step k. $k = 0$ means at the very beginning.

At the beginning, let the numbers be $a_1, a_2, \ldots, a_n$. Then

$$P_0 = a_1 \times a_2 + a_2 \times a_3 + \cdots + a_{n-1} \times a_n + a_n \times a_1 > 0.$$

At step $k(\geq 0)$, if 4 neighbouring numbers a, b, c, and d satisfy $(a-d)(b-c) > 0$, that is, $ac + bd - (ab + cd) \leq -1$, after the operation, b and c exchange their positions, and

$$P_{k+1} - P_k = (ac + cb + bd) - (ab + bc + cd) = ac + bd - (ab + cd) \leq -1.$$

Since P_0 is finite, and $P_k > 0$ decreases with k, the operation must end at some k.

7. Let the sum of all real numbers in the array be S. Obviously,

$$s_1 + s_2 + \cdots + s_n = t_1 + t_2 + \cdots + t_n = S.$$

For $k = 0, 1, \ldots, n-1$, let the subscripts equal in the sense of modulo n. We have

$$\sum_{i=1}^{n} a_{i,i+k} = \sum_{i=1}^{n} (s_i - t_{i+k}) = \sum_{i=1}^{n} s_i - \sum_{i=1}^{n} t_{i+k} = \sum_{i=1}^{n} s_i - \sum_{i=1}^{n} t_i = 0.$$

So, in the n numbers $a_{1,1+k}, a_{2,2+k}, \ldots, a_{n,n+k}$, there are at most $n - 1$ positive numbers. Let $k = 0, 1, \ldots, n - 1$, and we see that there are at most $n(n - 1)$ positive numbers in $a_{ij}(i, j = 1, 2, \ldots, n)$.

On the other hand, take the following array:

0	0	$\cdots$	0
1	1	$\cdots$	1
$\cdots$	$\cdots$	$\cdots$	$\cdots$
1	1	$\cdots$	1

It is easy to check that $a_{1j} = s_1 - t_j = -(n-1), j = 1, 2, \ldots, n,$

$$a_{ij} = s_i - t_j = 1, i = 2, 3, \ldots, n, j = 1, 2, \ldots, n.$$

The number of positive numbers in a_{ij} $(i, j = 1, 2, \ldots, n)$ is $n(n-1)$.

Summing up, the maximal number of positive numbers in n^2 numbers $a_{ij}(i, j = 1, 2, \ldots, n)$ is $n(n-1)$.

Remark. Because the value of $a_{i,j}$ is affected by a total of $(2n-2)$ numbers of the ith row and the jth column in the array (note: the change of the number in the ith row and the jth column does not affect the value of $a_{i,j}$), so, first divide n^2 numbers $a_{i,j}$ into n groups by $i - j$ modulo n. In each group, take the n real numbers as a whole, and see that their sum is 0, so each group has at most $n - 1$ positive numbers.

In general, some objects may be mutually restrictive, and it is more effective to look at them from the perspective of the whole rather than looking at each of them in isolation.

Chapter 10

1. Let the original six-digit telephone number $\overline{abcdef}$ be upgraded to the eight-digit number $\overline{2a8bcdef}$. Then, $81 \times \overline{abcdef} = \overline{2a8bcdef}$.

   Denote $x = b \times 10^4 + c \times 10^3 + d \times 10^2 + e \times 10 + f$, then

   $$81 \times a \times 10^5 + 81x = 208 \times 10^5 + a \times 10^6 + x,$$

   Yielding $x = 1250 \times (208 - 71a)$.

   Since $0 \le x \le 10^5, 0 \le 1250 \times (208 - 71a) < 10^5$. Hence, $\frac{128}{71} < a \le \frac{208}{71}$. That is, $a = 2$, and

   $$x = 1250 \times (208 - 71 \times 2) = 82500.$$

   Thus, the original telephone number is 282500.

2. Let $f(x) = (1+x)(1+x^2)(1+x^3)\cdots(1+x^{2009})$, then the answer is equal to the sum of expansion coefficients of odd powers of x of $f(x)$. So, the answer is

   $$\frac{f(1) - f(-1)}{2} = \frac{2^{2009} - 0}{2} = 2^{2008}.$$

3. Rewrite the original equation as

   $$m^2 - (10n + 7)m + 25n^2 - 7n = 0.$$

   Then, the discriminant $\Delta = (10n + 7)^2 - 4(25n^2 - 7n) = 168n + 49$ should be a perfect square. We see that $168n + 49 = 7(24n + 7)$, so let $24n + 7 = 7(1 + 6k)^2$, and then, $n = 7k(3k + 1)/2$, where k is any positive integer. Moreover,

   $$m = \frac{10n + 7 \pm 7(1 + 6k)}{2} = \frac{35(3k^2 + k) + 7 \pm 7(1 + 6k)}{2}$$

   $$= 7 + \frac{105k^2 + 77k}{2} \quad \text{or} \quad \frac{7k(15k - 1)}{2}.$$

   This completes the proof.

4. For $n = 1$, $d^2(1) = 1$. For $n > 1$, $d(n) > 1$. Let $d(n) = p_1^{\alpha_1} p_2^{\alpha_2} \cdots p_k^{\alpha_k}$ be the standard prime factorization of $d(n)$, then $n = p_1^{2\alpha_1} p_2^{2\alpha_2} \cdots p_k^{2\alpha_k}$. So,

$$d(n) = (2\alpha_1 + 1)(2\alpha_2 + 1) \cdots (2\alpha_k + 1),$$

and

$$p_1^{\alpha_1} p_2^{\alpha_2} \cdots p_k^{\alpha_k} = (2\alpha_1 + 1)(2\alpha_2 + 1) \cdots (2\alpha_k + 1). \tag{1}$$

The right-hand side of (1) is odd, so are $p_1, p_2, \ldots, p_k$. Therefore,

$$p_i^{\alpha_i} \geq 3^{\alpha_i} = (1+2)^{\alpha_i} \geq 1 + 2\alpha_i \quad (i = 1, 2, \ldots, k),$$

where equalities hold if and only if $p_i = 3$, $\alpha_i = 1$.

So, (1) holds if and only if $k = 1, p_1 = 3, \alpha_1 = 1$, that is, $n = 3^2 = 9$.

Summing up, the positive integers n that satisfy $n = d^2(n)$ are 1 and 9 only.

5. Let $x = (\overline{a_k a_{k-1} \ldots a_0 \cdot b_1 b_2 \ldots})_2$ be the binary representation of x. Then for any $n \in \mathbb{N}^+$, we have

$$\lfloor 2^n x \rfloor = (\overline{a_k a_{k-1} \ldots a_0 b_1 b_2 \ldots b_n})_2 \equiv b_n \pmod{2}.$$

Considering that $b_n \in \{0, 1\}$, there must be $(-1)^{\lfloor 2^n x \rfloor} = (-1)^{b_n} = 1 - 2b_n$. Thus,

$$S = \sum_{n=1}^{\infty} \frac{1 - 2b_n}{2^n} = \sum_{n=1}^{\infty} \frac{1}{2^n} - 2 \cdot \sum_{n=1}^{\infty} \frac{b_n}{2^n} = 1 - 2 \cdot (\overline{0.b_1 b_2 \ldots})_2$$

$$= 1 - 2(x - \lfloor x \rfloor) = 1 + 2\lfloor x \rfloor - 2x.$$

6. Considering that

$$e^{ix} = \cos x + i\sin x, \; e^{iy} = \cos y + i\sin y, \; e^{iz} = \cos z + i\sin z,$$

$$e^{i(x+y+z)} = \cos(x + y + z) + i\sin(x + y + z),$$

and the given equations, we obtain

$$a e^{ix} + b e^{iy} + c e^{iz}$$

$$= (a\cos x + b\cos y + c\cos z) + i(a\sin x + b\sin y + c\sin z)$$

$$= \cos(x + y + z) + i\sin(x + y + z)$$

$$= e^{i(x+y+z)}.$$

Multiply both sides by $e^{-i(x+y+z)}$ to yield

$$Z = a e^{-i(y+z)} + b e^{-i(z+x)} + c e^{-i(x+y)} = 1.$$

Hence,

$$\begin{cases} a\cos(y + z) + b\cos(z + x) + c\cos(x + y) = \operatorname{Re} Z = 1, \\ a\sin(y + z) + b\sin(z + x) + c\sin(x + y) = -\operatorname{Im} Z = 0. \end{cases}$$

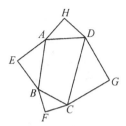

Figure 10(Ex).1

7. Let the points A, B, C, D, E, F, G, H correspond to complex numbers a, b, c, d, e, f, g, h, respectively.

Suppose without loss of generality that A, B, C, D are in the anti-clockwise order, as shown in Fig. 10(Ex).1. Then EG corresponds to the complex number

$$g - e = (g - d) + (d - a) + (a - e)$$

$$= \frac{\sqrt{2}}{2} e^{\frac{\pi i}{4}} (c - d) + (d - a) + \frac{\sqrt{2}}{2} e^{-\frac{\pi i}{4}} (a - b)$$

$$= \frac{1 + i}{2} (c - d) + (d - a) + \frac{1 - i}{2} (a - b)$$

$$= \frac{-a - b + c + d}{2} + \frac{-a + b + c - d}{2} i.$$

Similarly, FH corresponds to the complex number

$$h - f = \frac{-b - c + d + a}{2} + \frac{-b + c + d - a}{2} i.$$

So, $h - f = (g - e)i$, that is, EG is equal and vertical to FH.

8. Though this problem can be proved geometrically, we prefer to prove it analytically.

In the rectangular coordinate system with BC as x-axis and the perpendicular bisector of BC as y-axis, the coordinates of the three vertices of $\triangle ABC$ are $A(a, b)$, $B(-r, 0)$, and $C(r, 0)$, respectively, where r is the radius of the circle O, as shown in Fig. 10(Ex).2. Then, the equation of the circle O is

$$x^2 + y^2 = r^2.$$

Let the coordinates of the points P and Q be $P(x_1, y_1)$ and $Q(x_2, y_2)$, respectively. Then, the equations of the tangents AP and AQ are

$$xx_1 + yy_1 = r^2 \quad \text{and} \quad xx_2 + yy_2 = r^2,$$

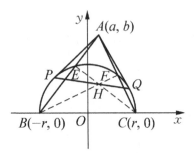

Figure 10(Ex).2

respectively. Since these two tangents both pass through point A, we see that

$$ax_1 + by_1 = r^2 \quad \text{and} \quad ax_2 + by_2 = r^2.$$

That is, points P and Q are both on the line $ax + by = r^2$, which is the equation of line PQ.

Let BF and CE be two altitudes of $\triangle ABC$. Since the slope of line AC is $\frac{b}{a-r}$, it follows that BF has slope $\frac{r-a}{b}$ and it is on the line

$$y = \frac{r - a}{b}(x + r).$$

That is,

$$(a - r)x + by = r^2 - ar. \tag{1}$$

Similarly, the equation of line CE is

$$(a + r)x + by = r^2 + ar. \tag{2}$$

By (1) plus (2), we obtain

$$ax + by = r^2. \tag{3}$$

That is, point H is on the line (3), the equation of line PQ. Hence, P, H, and Q are collinear.

9. Denote $x - \lfloor x \rfloor = \{x\}$. Let a and b be the binary representations of $\{\sqrt{2008}\}$ and $\{\sqrt{2009}\}$, respectively. Since $\sqrt{2008}$ and $\sqrt{2009}$ are irrational, a and b are not recurring decimals. The parity of f_n depends on whether the n-th decimal places of a and b are equal or not.

   Suppose, on the contrary, that there are finitely many odd numbers in $f_1, f_2, \ldots$. Then there exists $n \in \mathbb{N}^+$, such that $f_n, f_{n+1}, \ldots$ are all even. So $a - b$ is rational. A contradiction.

   Suppose on the contrary that there are finitely many even numbers in $f_1, f_2, \ldots$. Then there exists $n \in \mathbb{N}^+$, such that $f_n, f_{n+1}, \ldots$ are all odd numbers. Then all the digits of $a + b$ after the n-th decimal place are 1's, so $a + b$ is rational. A contradiction.

Chapter 11

1. Since $\sqrt{x^2 \pm x + 1} = \sqrt{(x \pm 1/2)^2 + (\sqrt{3}/2)^2}$, the inequality can be proved by considering $\triangle ABC$ with vertices $A(-1/2, 0)$, $B(1/2, 0)$, and $C(x, \sqrt{3}/2)$. By the fact that $|CA - CB| AB = 1$, we have (Fig. 11(Ex).1)

$$|\sqrt{x^2 + x + 1} - \sqrt{x^2 - x + 1}| < 1.$$

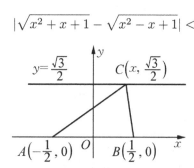

Figure 11(Ex).1

2. The number of ways of choosing 5 balls (each ball represents a person) in 10 balls is $\binom{10}{5} = 252$. That is the total number of ways of queuing.

   Let $p(n, m)$ denote the queuing ways of m persons each with a 5 Yuan note and n persons each with a 10 Yuan note such that no one runs short of change. Consider a lattice path from A to B: if a ticket is sold to a person with a 5 Yuan note, then move up 1 unit; if sold to a person with a 10 Yuan note, then move 1 unit right. Obviously, no one runs short of change if and only if the path is in the upper isosceles right-angled triangle with hypotenuse AB as shown in Fig. 11(Ex).2. Then $p(0, m) = 1$, for $m \in \mathbb{N}^+$; $p(n, m) = 0$, for $m < n$.

235

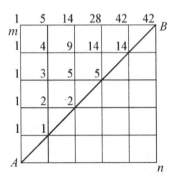

Figure 11(Ex).2

Since there are two ways to arrive at point (n, m): from $(n-1, m)$ or $(n, m-1)$, we have $p(n, m) = p(n-1, m) + p(n, m-1)$ for $m \geq n \geq 1$ (as shown in Fig. 11(Ex).2, $p(n, m)$ is shown at grid (n, m)).

We can check that $p(n, m) = \frac{(m+n)!(m+1-n)}{n!(m+1)!}$ by induction, so, $p(5, 5) = 42$, the probability is $42/252 = 1/6$.

3. By Viète's theorem, we have $\begin{cases} \alpha + \beta = -z_1, \\ \alpha\beta = z_2 + m. \end{cases}$ So,

$$(\alpha - \beta)^2 = (\alpha + \beta)^2 - 4\alpha\beta = z_1^2 - 4z_2 - 4m.$$

Thus,

$$|(\alpha - \beta)^2| = |z_1^2 - 4z_2 - 4m| = |16 + 20i - 4m| = 4|m - (4 + 5i)|.$$

On the other hand, $|(\alpha - \beta)^2| = |\alpha - \beta|^2 = 28$. So, $|m - (4 + 5i)| = 7$. That is, m is on the circle K with centre $A(4, 5)$ and radius 7.

Since $|OA| = \sqrt{4^2 + 5^2} = \sqrt{41} < 7$, the origin O is in the circle K. Let the extension of OA meet the circle K at point B, and the extension of AO meet the circle K at point C. Then,

$$|m|_{\max} = |OB| = \sqrt{41} + 7, \quad |m|_{\min} = |OC| = 7 - \sqrt{41}.$$

4. The graph of $f(x)$ and the line $y = kx (k > 0)$ do not intersect for $x \in (-\infty, 0)$ and intersect at exactly two points for $x \in [0, \pi]$. So, the third intersection point must be a tangent point $(x, y) = (\alpha, -\sin\alpha), \alpha \in \left(\pi, \frac{3\pi}{2}\right)$. Since $f'(\alpha) = (-\sin\alpha) = -\cos\alpha = \frac{-\sin\alpha}{\alpha}$, that is, $\alpha = \tan\alpha$, therefore,

$$\frac{\cos\alpha}{\sin\alpha + \sin 3\alpha} = \frac{\cos\alpha}{2\sin 2\alpha \cos\alpha} = \frac{1}{2\sin 2\alpha} = \frac{\cos^2\alpha + \sin^2\alpha}{4\sin\alpha\cos\alpha}$$

$$= \frac{1 + \tan^2\alpha}{4\tan\alpha} = \frac{1 + \alpha^2}{4\alpha}.$$

5. Make $\triangle ABC$ with $AB = 5, AC = 4, BC = 3, \angle ACB = 90°$. Make an equilateral triangle $\triangle ACD$ outside $\triangle ABC$. The circumcircle of $\triangle ACD$ meets the circle with diameter BC at two points C and O. Then by the plane geometry, $\angle BOC = 90°, \angle AOC = 120°$, and $\angle AOB = 150°$.

Let $AO = x, BO = \frac{y}{\sqrt{3}}, CO = z$. By $S_{\triangle AOB} + S_{\triangle BOC} + S_{\triangle COA} = S_{\triangle ABC}$, we have

$$\frac{1}{2}x \cdot \frac{y}{\sqrt{3}} \sin 150° + \frac{1}{2}z \cdot \frac{y}{\sqrt{3}} + \frac{1}{2}x \cdot z \cdot \sin 120° = \frac{1}{2} \times 3 \times 4,$$

which can be simplified to $xy + 2yz + 3xz = 24\sqrt{3}$.

6. Let $A, B,$ and C be points corresponding to the complex numbers z_1, z_2 and xz_2 $(0 \leq x \leq 1)$ in the complex plane, respectively, as shown in Fig. 11(Ex).3. Obviously, point C is on the segment OB, vector $\overrightarrow{BA}$ corresponds to the complex number $z_1 - z_2$, vector $\overrightarrow{CA}$ corresponds to the complex number $z_1 - xz_2$. By $|z_1| \leq \alpha|z_1 - z_2|, |\overrightarrow{OA}| \leq \alpha|\overrightarrow{BA}|$. Hence,

$$|z_1 - xz_2|_{\max} = |\overrightarrow{AC}|_{\max} = \max\{|\overrightarrow{OA}|, |\overrightarrow{BA}|\} = \max\{|z_1|, |z_1 - z_2|\}$$

$$= \max\{\max|z_1|, |z_1 - z_2|\} = \max\{\alpha|z_1 - z_2|, |z_1 - z_2|\}.$$

Thus, $\lambda(\alpha) = \max\{\alpha, 1\}$.

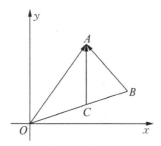

Figure 11(Ex).3

7. Let, say, $x_1 > x_2 > x_3,$ $a \geq b \geq c$ (the two equalities do not hold simultaneously). By Chebyshev's inequality, we have

$$\overrightarrow{OA} \cdot \overrightarrow{OB_1} = x_1a + x_2b + x_3c > \frac{1}{3}(a + b + c)(x_1 + x_2 + x_3) = 0.$$

So, $\cos \angle AOB_1 > 0$. Furthermore, O, A, B_1 are not collinear, thus $\angle AOB_1$ is acute.

On the other hand, by Chebyshev's inequality, we have

$$x_3 a + x_2 b + x_1 c < \frac{1}{3}(a+b+c)(x_3 + x_2 + x_1) = 0.$$

Hence,

$$\max \overrightarrow{OA} \cdot \overrightarrow{OB_2}, \overrightarrow{OA} \cdot \overrightarrow{OB_3}$$

$$= \max x_1 a + x_3 b + x_2 c, x_2 a + x_1 b + x_3 c$$

$$\geq \frac{1}{2}(x_1 a + x_3 b + x_2 c) + (x_2 a + x_1 b + x_3 c)$$

$$= -\frac{1}{2}(x_3 a + x_2 b + x_1 c) > 0.$$

Therefore, at least one of the two angles $\angle AOB_2$ and $\angle AOB_3$ is acute. Summing up, at least two angles of $\angle AOB_i (i = 1, 2, \ldots, 6)$ are acute.

8. Let $P, A, B,$ and C be points corresponding to complex numbers $z_0, z_1, z_2,$ and z_3 in the complex plane, respectively. Let

$$f(z) = \frac{(z - z_2)(z - z_3)}{(z_1 - z_2)(z_1 - z_3)} + \frac{(z - z_3)(z - z_1)}{(z_2 - z_3)(z_2 - z_1)} + \frac{(z - z_1)(z - z_2)}{(z_3 - z_1)(z_3 - z_2)},$$

$$g(z) = \frac{(z - z_1)^2}{(z_1 - z_2)(z_1 - z_3)} + \frac{(z - z_2)^2}{(z_2 - z_3)(z_2 - z_1)} + \frac{(z - z_3)^2}{(z_3 - z_1)(z_3 - z_2)}.$$

It is easy to check that $f(z_i) = g(z_i) = 1$, $i = 1, 2, 3$. That is, $f(z) - 1 = 0$ and $g(z) - 1 = 0$ each has at least three complex roots. But $f(z) - 1$ and $g(z) - 1$ are all polynomials with order less than 3, so, $f(z) \equiv 1, g(z) \equiv 1$.

Since

$$1 = f(z_0) \leq \frac{|z_0 - z_2| \cdot |z_0 - z_3|}{|z_1 - z_2| \cdot |z_1 - z_3|} + \frac{|z_0 - z_3| \cdot |z_0 - z_1|}{|z_2 - z_3| \cdot |z_2 - z_1|}$$

$$+ \frac{|z_0 - z_1| \cdot |z_0 - z_2|}{|z_3 - z_1| \cdot |z_3 - z_2|}$$

$$= \frac{PB \cdot PC}{cb} + \frac{PC \cdot PA}{ac} + \frac{PA \cdot PB}{ba},$$

we see that $a \cdot PB \cdot PC + b \cdot PC \cdot PA + c \cdot PA \cdot PB \geq abc$, that is, (a) holds.

Similarly, since

$$1 = g(z_0) \leq \frac{|z_0 - z_1|^2}{|z_1 - z_2| \cdot |z_1 - z_3|} + \frac{|z_0 - z_2|^2}{|z_2 - z_3| \cdot |z_2 - z_1|}$$

$$+ \frac{|z_0 - z_3|^2}{|z_3 - z_1| \cdot |z_3 - z_2|}$$

$$= \frac{PA^2}{cb} + \frac{PB^2}{ac} + \frac{PC^2}{ba},$$

we see that $a \cdot PA^2 + b \cdot PB^2 + c \cdot PC^2 \geq abc$, that is, (b) holds.

Chapter 12

1. Obviously, the number of intersection points in the circle is 0 if $n \leq 3$. For $n \geq 4$, regard 4 points A_1, A_2, A_3, and A_4 on the circle as the vertices of a convex quadrilateral, of which two diagonals $A_1 A_3$ and $A_2 A_4$ meet at a point P in the circle. Since no three chords meet at one point in the circle, the set $\{A_1, A_2, A_3, A_4\}$ of four points on the circle one-to-one corresponds to a point P in the circle. So, the number of intersection points of all chords is the number of ways of taking four points from n points on the circle. That is $\binom{n}{4}$.

2. Denote the *alternating sum* of the set $A \subseteq \{1, 2, \ldots, n\}$ by $f(A)$. Here, we set $f(\emptyset) = 0$.

   For each $X \subseteq \{1, 2, \ldots, n-1\}$, let $Y = X \cup \{n\} = \{a_1, a_2, \ldots, a_k\}$, which is a subset of $\{1, 2, \ldots, n\}$ containing n, where $n = a_1 > a_2 > \cdots > a_k \geq 1, k \in \mathbb{N}^+$. Then for pair sets (X, Y), we have

$$f(Y) + f(X) = \sum_{i=1}^{k} (-1)^{i-1} a_i + \sum_{j=2}^{k} (-1)^j a_j$$

$$= n + \sum_{i=2}^{k} (-1)^{i-1} a_i - \sum_{j=2}^{k} (-1)^{j-1} a_j = n.$$

   Since all subsets of $\{1, 2, \ldots, n\}$ can be divided into 2^{n-1} distinct pair sets (X, Y), the total sum of the *alternating sums* is $n \cdot 2^{n-1}$.

3. There are $\binom{n}{r}$ subsets of r elements of $P = \{1, 2, \ldots, n\}$. First, we find the sum of the minimums of these subsets. For any set $A \subset P$ with r elements and minimal element a, let $Q = P \cup \{0\}$, and let A correspond

to a subsets of Q, $A \cup \{0\}$, $A \cup \{1\}, \ldots, A \cup \{a-1\}$, each with $r+1$ elements. Conversely, every subset of Q corresponds to a unique subset of P (by eliminating its minimal element). Therefore, the sum of all minimums of the subsets of r elements of P is the number of subsets with $(r+1)$ elements of Q, that is, $\binom{n+1}{r+1}$.

Therefore, $f(r,n) = \dfrac{\binom{n+1}{r+1}}{\binom{n}{r}} = \frac{n+1}{r+1}$.

4. (a) Let A be an *odd subset* of $\{1, 2, \ldots, n\}$. Define a mapping

$$f : \begin{cases} A \mapsto B = A\backslash\{1\}, & \text{if } 1 \in A, \\ A \mapsto B = A \cup \{1\}, & \text{if } 1 \notin A, \end{cases}$$

which is injective.

The mapping f is also surjective, since if $1 \in B$, then $A = B\backslash\{1\}$ and if $1 \notin B$, then $A = B \cup \{1\}$. Thus, the mapping f is bijective. Therefore, the number of *odd subsets* and the number of *even subsets* are equal, and they are $2^n/2 = 2^{n-1}$.

(b) If P is a set of finite sets A_i of finitely many real numbers, define $m(P) = \sum_{A_i \in P} \sum_{x \in A_i} x$, and call it the sum to P.

For $X = \{1, 2, \ldots, n\}$, $(n \geq 3)$, denote all *odd subsets* and *even subsets* containing $\{n, (n-1)\}$ by S_1 and T_1, respectively. Denote all *odd subsets* and *even subsets* containing n but not $(n-1)$ by S_2 and T_2, respectively. Denote all *odd subsets* and *even subsets* containing $(n-1)$ but not n by S_3 and T_3, respectively. Denote all *odd subsets* and *even subsets* not containing n or $(n-1)$ by S_4 and T_4, respectively.

As a corollary of (a), the number of subsets in S_i and the number of subsets in T_i are all equal to $2^{(n-2)-1} = 2^{n-3}$, $i = 1, 2, 3, 4$.

We see that the set difference of each set in S_1 and $\{n, n-1\}$ corresponds to a set in T_4, thus,

$$m(S_1) = m(T_4) + (2n-1)2^{n-3}.$$

Similarly,

$$m(S_4) = m(T_1) - (2n-1)2^{n-3},$$
$$m(S_2) = m(T_3) + 2^{n-3},$$
$$m(S_3) = m(T_2) - 2^{n-3}.$$

Summing up the above four equations yields

$$m(S_1 \cup S_2 \cup S_3 \cup S_4) = m(T_1 \cup T_2 \cup T_3 \cup T_4).$$

Note that $m(X) = (1+2+\cdots+n)2^{n-1} = n(n+1)2^{n-2}$, since each $i \in X$ appears in 2^{n-1} subsets. So, the total sum of elements of *odd subsets* is the half of $m(X)$, that is $n(n+1)2^{2n-3}$.

5. Denote $S = 1 + 2 + \cdots + n = \frac{n(n+1)}{2}$. If $n \equiv 1 \pmod 4$, S is odd. For a given permutation $P = (a_1, a_2, \ldots, a_n)$ of $(1, 2, \ldots, n)$, by the definition of k_P, we see that

$$a_1 + a_2 + \cdots + a_{k_P} < a_{k_P+1} + a_{k_P+2} + \cdots + a_n,$$

$$a_1 + a_2 + \cdots + a_{k_P+1} \geq a_{k_P+2} + a_{k_P+3} + \cdots + a_n.$$

Note that $(a_1 + a_2 + \cdots + a_{k_P+1}) + (a_{k_P+2} + a_{k_P+3} + \cdots + a_n) = 1 + 2 + \cdots + n = S$ is odd, the equality cannot hold.

Take the reverse order $P' = (b_1, b_2, \ldots, b_n) = (a_n, a_{n-1}, \ldots, a_1)$ of P. Similarly, we see that

$$b_1 + b_2 + \cdots + b_{n-k_P-1} < b_{n-k_p} + b_{n-k_P+1} + \cdots + b_n,$$

$$b_1 + b_2 + \cdots + b_{n-k_P} < b_{n-k_p+1} + b_{n-k_P+2} + \cdots + b_n.$$

Thus, $k_{P'} = n - k_P - 1$, that is, for permutations P and P' in reverse order of each other, $k_P + k_{P'} = n - 1$.

Note that the $n!$ permutations of $(1, 2, \ldots, n)$ can be divided into $n!/2$ pairs of permutations in reverse order of each other. So, the sum of all k_P's is equal to $\frac{n!}{2} \cdot (n-1)$.

6. Let $\gcd(a, n) = 1$, and $1 \leq a < n/2$, then $\gcd(n-a, n) = 1$. We note that $a \neq n - a$, otherwise, $n = 2a$, since $n > 2$, $a > 1$, leads to $\gcd(a, n) \neq 1$. Thus, we can match a and $n - a$ as a pair $(a, n - a)$. Since

$$a^3 + (n-a)^3 = n[a^2 - a(n-a) + (n-a)^2]$$

is a multiple of n, summing up the above equalities for all such a, the conclusion follows.

7. We show that, for $k = 1, 2, \ldots, \frac{p-1}{2}$,

$$k^{k^2-k+1} + (p-k)^{(p-k)^2-(p-k)+1} \equiv 0 \pmod p. \tag{1}$$

In fact,

$$(p-k)^2 - (p-k) + 1 = k^2 - k + 1 + (p-2k)(p-1) \equiv k^2 - k + 1 (\bmod\ p-1).$$

And note that $(p-k,p) = 1$. By Fermat's little theorem, $(p-k)^{p-1} \equiv 1 (\bmod\ p)$, so,

$$\begin{aligned}
(p-k)^{(p-k)^2-(p-k)+1} \equiv (p-k)^{k^2-k+1} &\equiv (-k)^{k^2-k+1} \\
&= (-1)^{k(k-1)+1} k^{k^2-k+1} \\
&\equiv -k^{k^2-k+1} (\bmod\ p).
\end{aligned}$$

That is, (1) holds.
Take sum in (1) for $k = 1, 2, \ldots, \frac{p-1}{2}$:

$$\sum_{k=1}^{p-1} k^{k^2-k+1} = \sum_{k=1}^{\frac{p-1}{2}} \left(k^{k^2-k+1} + (p-k)^{(p-k)^2-(p-k)+1} \right) \equiv 0 \ (\bmod\ p).$$

Thus, $\sum_{k=1}^{p-1} k^{k^2-k+1}$ is divisible by p.

8. Take the radical conjugate $B = k + \frac{1}{2} - \sqrt{k^2 + \frac{1}{4}}$ of A, then $A + B = 2k + 1$, $AB = k$. So, A and B are the two real roots of the equation $x^2 - (2k+1)x + k = 0$. Then, for any positive integer $n \geq 3$, we have

$$A^n - (2k+1)A^{n-1} + kA^{n-2} = 0,$$
$$B^n - (2k+1)B^{n-1} + kB^{n-2} = 0.$$

Denote $a_n = A^n + B^n - 1$, and sum up the above equations to yield

$$(a_n + 1) - (2k+1)(a_{n-1} + 1) + k(a_{n-2} + 1) = 0.$$

That is,

$$a_n = (2k+1)a_{n-1} - ka_{n-2} + k. \tag{1}$$

Note that $a_1 = A + B - 1 = 2k$, $a_2 = A^2 + B^2 - 1 = 4k^2 + 2k$ are all divisible by k, we see from (1), for all positive integers n, a_n is divisible by k. Further, $0 < B < 1$, so,

$$a_n = \lfloor a_n \rfloor = \lfloor A^n + B^n - 1 \rfloor = \lfloor A^n \rfloor.$$

Hence, $\lfloor A^n \rfloor$ is divisible by k.

9. Denote $S_{i,j}$ the set of points (x, i, j) in S. That is, the set of points in S whose projection on the yz-plane has coordinates (i, j).

Obviously, $S = \bigcup_{(i,j) \in S_1} S_{i,j}$. By Cauchy's inequality,

$$|S|^2 = \left(\sum_{(i,j) \in S_1} |S_{i,j}| \right)^2 \leq \sum_{(i,j) \in S_1} 1^2 \times \sum_{(i,j) \in S_1} |S_{i,j}|^2 = |S_1| \sum_{(i,j) \in S_1} |S_{i,j}|^2.$$

Construct the set $X = \bigcup_{(i,j) \in S_1} (S_{i,j} \times S_{i,j})$, and $|X| = |\sum_{(i,j) \in S_1} S_{i,j}|^2$.

Then define the mapping:

$$f : X \to S_2 \times S_3 \text{ by } f((x, i, j), (x', i, j)) = ((x, j), (x', i)).$$

f is obviously injective, hence, $|X| \leq |S_2 \cdot S_3|$. Consequently,

$$|S|^2 \leq |S_1| \cdot |X| \leq |S_1| \cdot |S_2| \cdot |S_3|.$$

Chapter 13

1. Let the answer be a_n for n people. Then it is easy to see that $a_1 = 1, a_2 = 2$.

   If $n \geq 3$, consider the position of the nth person at another desk, and there are two possibilities:

   If the nth person still queues up at the nth position, then the number of ways the first $(n-1)$ persons queue up is a_{n-1};

   If the nth person queues up at the $(n-1)$th position, then the $(n-1)$th person must queue up at the nth position, and the number of ways of the first $(n-2)$ persons is a_{n-2}.

   Thus, let $a_0 = 1$, and we have the recurrence relation $a_{n+2} = a_{n+1} + a_n (n \geq 0)$ of the Fibonacci sequence.

   The formula is $a_n = \frac{((\sqrt{5}+1)/2)^{n+1} - (-(\sqrt{5}-1)/2)^{n+1}}{\sqrt{5}}$, and $a_{12} = 233$.

2. We establish the relation between A_{n+1} and A_n. If the nth and the $(n+1)$th throws are all by A, the probability is $A_n/6$. If the nth throw is by B (the probability is $1 - A_n$) and the $(n+1)$th throw is by A, the probability is $\frac{5}{6}(1 - A_n)$. Thus,

$$A_{n+1} = \frac{1}{6}A_n + \frac{5}{6}(1 - A_n) = \frac{5}{6} - \frac{2}{3}A_n, \quad \text{with } A_1 = 1.$$

   We have $A_n = \frac{1}{2} + \frac{1}{2}\left(-\frac{2}{3}\right)^{n-1}$.

3. Such number of n digits can be classified into two classes.

   (a) Digit 5 followed by $(n-1)$ digits without 5, there are 4^{n-1} such numbers;

   (b) An $(n-1)$-digit number containing 5, and no digit 3 ahead of 5, then put number 1, 2, 4, or 5 ahead of the $(n-1)$ digits. There are $4f(n-1)$ such numbers.

Hence,

$$f(n) = 4f(n-1) + 4^{n-1}, \quad f(1) = 1,$$

$$\frac{f(n)}{4^n} = \frac{f(n-1)}{4^{n-1}} + \frac{1}{4},$$

$$\frac{f(n)}{4^n} = \frac{1}{4} + \frac{1}{4}(n-1) = \frac{1}{4}n.$$

Therefore, $f(n) = n \cdot 4^{n-1}$.

Remark. This problem can also be solved by other simple methods.

4. Denote $P_k(x) = ((\cdots(((x-2)^2 - 2)^2 - 2)^2 - \cdots - 2)^2 - 2)^2$ (totally k pairs of parentheses), of which the constant term is

$$P(0) = ((\cdots(((0-2)^2, -2)^2 - 2)^2 - \cdots - 2)^2 - 2)^2 = 4.$$

Denote the coefficients of the first- and the second-order terms of $P_k(x)$ by A_k and B_k, respectively. Since $P_1(x) = (x-2)^2$, we see $A_1 = -4$, $B_1 = 1$.

Let us find the recurrence relations for A_k and B_k.

On the one hand, $P_k(x) = 4 + A_k x + B_k x^2 + O(x)$ (where $O(x)$ is a polynomial only with terms higher than second order). If $k \geq 2$,

$$P_k(x) = (P_{k-1}(x) - 2)^2 = (4 + A_{k-1}x + B_{k-1}x^2 + O(x) - 2)^2$$

$$= 4 + 4A_{k-1}x + (4B_{k-1} + A_{k-1}^2)x^2 + O(x),$$

Compare the coefficients of x, x^2 in the two expansions of $P_k(x)$, to obtain

$$A_k = 4A_{k-1}, \quad B_k = 4B_{k-1} + A_{k-1}^2 (k \geq 2).$$

So, first, we find $A_n = 4^{n-1}A_1 = -4^n$. Then, we change the second recurrence relation into $\frac{B_k}{4^k} = \frac{B_{k-1}}{4^{k-1}} + \frac{A_{k-1}^2}{4^k}$, sum up for $k = 2, 3, \ldots, n$, yielding

$$\frac{B_n}{4^n} = \frac{B_1}{4} + \frac{A_{n-1}^2}{4^n} + \frac{A_{n-2}^2}{4^{n-1}} + \cdots + \frac{A_1^2}{4^2}$$

$$= \frac{1}{4} + 4^{n-2} + 4^{n-3} + \cdots + 4^0 = \frac{1}{4} \cdot \frac{4^n - 1}{3}.$$

Therefore, the coefficient of x^2 in $P_n(x)$ is $B_n = \frac{4^{n-1}(4^n-1)}{3}$.

5. Discuss the following general problem. For $n \geq 3$, select n terms from four sequences $A = a_1, a_2, \ldots, a_n$, $B = b_1, b_2, \ldots, b_n$, $C = c_1, c_2, \ldots, c_n$, $D = d_1, d_2, \ldots, d_n$, such that:

   (I) Each subscript $1, 2, \ldots, n$ appears;
   (II) Two terms with neighbouring subscripts are not in the same sequence (subscripts n and 1 are neighbours).

   Let x_n be the number of ways of selection. Divide a disc into n sectors, labelled $1, 2, \ldots, n$ in order. Colour four sequences A, B, C, D in red, black, blue, and green, respectively. For any selection, colour the ith sector in the colour which it comes from. Then x_n is the number of ways of colouring the n sectors of the disc in four colours such that neighbouring sectors are in different colours. It is easy to see, $x_1 = 4$ and $x_2 = 12$. Colour the first sector in any colour and each following sector in a colour different from that of the previous sector. There are $x_n + x_{n-1} = 4 \cdot 3^{n-1}$ ways. On the other hand, (a) if the nth sector and the first have different colours, it is a valid selection for n sectors; (b) if the nth and the first sector have the same colour, then the $(n-1)$th has a different colour and it is a valid selection for $(n-1)$ sectors. And moreover, the correspondence is one-to-one. Hence,

$$-1^n x_n - -1^{n-1} x_{n-1} = -4 \cdot -3^{n-1}.$$

$$-1^{n-1} x_{n-1} - -1^{n-2} x_{n-2} = -4 \cdot -3^{n-2};$$

$$\cdots$$

$$-1^3 x_3 - -1^2 x_2 = -4 \cdot -3^2;$$

$$-1^2 x_2 = -4 \cdot -3.$$

Summing up, we have $-1^n x_n = -3^n + 3$, that is, $x_n = 3^n + 3 \cdot -1^n (n \geq 2)$. Therefore, the answer is $x_{13} = 3^{13} - 3$.

Remark. If the four colours are generalized to m colours, the answer for n sectors is

$$x_n = (m-1)^n + (m-1)(-1)^n.$$

6. Obviously, there is no maximal element in S, so S is an infinite set. Denote

$$A_k = \{x \in \mathbb{N}^+ \mid x \equiv k \ (\mathrm{mod}\ 7)\} \quad (k = 0, 1, \ldots, 6).$$

And denote the proposition '$S \cap A_k$ is an infinite set' by P_k. Obviously, P_0 is what we should prove.

First, we prove a lemma: If P_k is true, then P_{k^2+3} is true (the subscript is modulo 7).

Proof of the Lemma. If P_k is true, then there exists a sequence of positive integers $a_1 < a_2 < \cdots < a_n < \cdots$, satisfying

$$a_1 \equiv a_2 \equiv \cdots \equiv k \,(\mathrm{mod}\ 7).$$

Then,

$$a_1 a_i + 3 \equiv k^2 + 3 \,(\mathrm{mod}\ 7), \quad a_1 a_i + 3 < a_1 a_{i+1} + 3 \ (i \in \mathbb{N}^+),$$

so $S \cap A_{k^2+3}$ contains infinitely many elements, that is, P_{k^2+3} is true.

Now return to the original problem. Since S is an infinite set, by the drawer principle, one of the propositions $P_0, P_1, \ldots, P_6$ is true. By using the lemma repeatedly, we have $P_1 \Rightarrow P_4 \Rightarrow P_5 \Rightarrow P_0$, while $P_2 \Rightarrow P_0, P_3 \Rightarrow P_5, P_6 \Rightarrow P_4$, so P_0 is always true. That is, there are infinitely many numbers in S divisible by 7.

7. Denote the number of $2 \times n$ grids with *stranded* black squares by a_n and the number of $2 \times n$ grids with no *stranded* black squares by b_n. Then $b_n = 2^{2n} - a_n$ with $a_1 = 0$.

By definition, *stranded* black squares can only be on the second row.

(I) If there are *stranded* squares in the first $(n-1)$ squares of the second row, then, no matter how the nth column is coloured, there are *stranded* black squares in the $2 \times n$ grids. There are totally $2^2 a_{n-1} = 4a_{n-1}$ cases.

(II) If there are no *stranded* black squares in the first $(n-1)$ squares of the second row, then the square at the nth column and the second row must be the unique *stranded* black square of the $2 \times n$ grid. Then the square above it is white, and there are white squares in the first $n-1$ squares of the second row. If the first $n-1$ squares of the second row are all black, for any colouring of the first row, there do not exist *stranded* black squares. So, these 2^{n-1} cases should be excluded. So, there are $b_{n-1} - 2^{n-1} = 2^{2(n-1)} - a_{n-1} - 2^{n-1}$ cases.

Summing up (I) and (II), we see that

$$a_n = 4a_{n-1} + 2^{2(n-1)} - a_{n-1} - 2^{n-1} = 3a_{n-1} + 4^{n-1} - 2^{n-1}.$$

That is,

$$\frac{a_n}{3^{n-1}} = \frac{a_{n-1}}{3^{n-2}} + \left(\frac{4}{3}\right)^{n-1} - \left(\frac{2}{3}\right).$$

Substitute n in the above equality with $2, 3, \ldots, n$ and sum up. Note that $a_1 = 0$, we find that

$$\frac{a_n}{3^{n-1}} = \sum_{k=2}^{n} \left(\frac{4}{3}\right)^{n-1} - \sum_{k=2}^{n} \left(\frac{2}{3}\right)^{n-1}$$

$$= 4 \times \left(\left(\frac{4}{3}\right)^{n-1} - 1\right) - 2 \times \left(1 - \left(\frac{2}{3}\right)^{n-1}\right)$$

$$= \frac{4^n + 2^n}{3^{n-1}} - 6.$$

That is, $a_n = 4^n + 2^n - 6 \times 3^{n-1} = 4^n + 2^n - 2 \times 3^n$. Therefore, the answer is

$$b_n = 2^{2n} - a_n = 2 \times 3^n - 2^n.$$

8. First, we can check that $n = 1, 2, 5, 6$, the first player cannot win; $n = 3, 4, 7, 8$, the first player can win. So, we guess that if and only if $n \equiv 0, 3 \pmod 4$, the first player has a winning strategy.

   We use an unordered quad P of positive integers to represent the numbers of matches in the four piles. And denote the sets of all states that the first player wins or loses by W and L, respectively.

   We first prove a lemma: let an unordered quad of positive integers be written in the form of $P = (2^{n_1} a_1, 2^{n_2} a_2, 2^{n_3} a_3, 2^{n_4} a_4)$ (where $n_i \in \mathbb{N}$, a_i being odd, $i = 1, 2, 3, 4$), if there are distinct i, j, k, such that $n_i = n_j < n_k$ (we call it case 1), then $P \in W$; if $n_1 = n_2 = n_3 = n_4$ (we call it case 2), then $P \in L$.

   In fact, in case 1, the first player takes the pile not labelled i, j, k, and divides the pile k into two piles of 2^{n_i} and $2^{n_i}(2^{n_k - n_i} a_k - 1)$ matches. Then, $a_i, a_j, 1, 2^{n_k - n_i} a_k - 1$ are all odd. It is in case 2.

   In case 2, let $n_1 = n_2 = n_3 = n_4 = n$. The first player takes pile 1, and divides pile 2 into two piles of $2^{m_1} b_1$ and $2^{m_2} b_2$ matches ($m_1, m_2 \in \mathbb{N}, b_1, b_2$ are odd). Considering a_2 is odd, and $2^{m_1} b_1 + 2^{m_2} b_2 = 2^n a_2$, so, either $m_1 = m_2 < n$, or $m_1 = n < m_2$, or $m_2 = n < m_1$, anyway, it is in case 1.

   In case 1, the game can always go on; since the game ends in finite steps, the first player wins. Now the lemma is proved.

   In the following, we show that: for any $k, l \in \mathbb{N}^+$, if $P = (2, 2, 4k, l)$, $(2, 2, 4k - 1, l)$ (we call it case 3), then $P \in W$; if $P = (2, 2, 2, 4k - 2)$, $(2, 2, 2, 4k - 3)$ (we call it case 4), then $P \in L$.

In fact, in the case 3, the first player takes out pile 4, and divides pile 3 into two piles of 2 and $4k - 2$ matches, or 2 and $4k - 3$ matches, it is the case 4.

In case 4, we show that any operation will result in a win state or a state in case 3:

If one divides some pile of the first three piles, then it becomes quad $(1, 1, 2, m)$. By the lemma, we know $(1, 1, 2, m) \in W$. If one takes away some pile in the first three piles and divides pile 4, then it becomes quad $(2, 2, c, d)$. If any one of c, d is $\equiv 3 \pmod 4$, then it is in case 3. If c and d are both $\equiv 1$ or $2 \pmod 4$, then it must be $c, d \equiv 1 \pmod 4$ since $c + d \equiv 1, 2 \pmod 4$. By the lemma, $(2, 2, c, d) \in W$.

The conclusion follows by recursion. By the above conclusion, it is easy to see that if and only if $n \equiv 0, 3 \pmod 4$, the first player has a winning strategy.

Chapter 14

1. The answer is negative. Colour the 8×8 grid in black and white as a chessboard, then cut two squares at the opposite corners. If the answer is positive, then the 31 dominoes cover 31 black squares and 31 white squares. But the numbers of white and black squares are different on the board, 30 in one colour, and 32 in the other colour.

2. The answer is negative. Divide the cube into 27 small cubes (each represents a room) and colour them black or white such that the centre cube is white and any two neighbouring cubes are in different colours as shown in Fig. 14(Ex).1. If the answer is positive, then the beetle starts walking from the middle white cube, and continues to black, white, black, white, and so on. At the end, the beetle passes 14 white cubes and 13 black cubes. Since there are 14 black cubes and 13 white cubes, this is a contradiction.

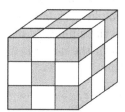

Figure 14(Ex).1

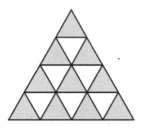

Figure 14(Ex).2

3. Colour the n^2 small equilateral triangles in black and white as shown in Fig. 14(Ex).2. The number of black triangles is

$$1 + 2 + 3 + \cdots + n = \frac{1}{2}n(n+1).$$

And the number of white triangles is

$$1 + 2 + 3 + \cdots + (n-1) = \frac{1}{2}n(n-1).$$

Since two triangles with consecutive labelling numbers have a common side and so have different colours, hence, the number of black labelled triangles is at most one more than that of the white labelled triangles. Therefore,

$$m \leq 2 \times \frac{1}{2}n\,(n-1) + 1 = n^2 - n + 1.$$

4. Colour these triangles in black and white such that any two triangles with a common side are in different colours. Since each vertex of the convex n-gon is the common vertex of an odd number of triangles, the n sides of the n-gon are all in a colour (say, in black as shown in Fig. 14(Ex).3). So, the difference between the number of black angles and the number of white angles is n. That is, if the n-gon is divided into x black triangles and y white triangles, then

$$n = 3x - 3y = 3(x - y),$$

n is a multiple of 3.

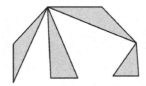

Figure 14(Ex).3

5. Denote each person as a point: if two persons know each other, link the corresponding points in blue line; if not, in red line. So, the problem is changed to a graph edge colouring problem:

   Colour the edges of a complete 9-graph K^9 in red and blue. If there is no red triangle in the graph, prove that there exists a complete 4-graph K^4 with all edges in blue.

   Considering the eight edges with a common vertex A, there are three cases:

   (1) There are four edges in red, say AB_i, $(1 \leq i \leq 4)$. Since there is no red triangle, the four vertices B_1, B_2, B_3, B_4 are all linked by blue edges.

   (2) There are six edges in blue. Consider the six vertices linked to A. They consist of a K^6, in which there is a blue triangle by the Ramsey theorem (assume there is no red triangle). So, a blue K^4 is present with vertex A and three vertices of the blue triangle.

   (3) There are three edges in red and five edges in blue. If this is the case for each vertex, then the total number of red edges is $3 \times 9/2 = 13.5$, this implies that one vertex must have four or more red edges, or six or more blue edges, as in case (1) or (2).

6. Denote the six points by $A_1, A_2, \ldots, A_6$. Colour those edges $A_i A_j$ $(1 \leq i < j \leq 6)$ in red if it can be the longest side of a triangle, in blue otherwise. That is a two-colour complete graph K^6. So, there is no blue triangle in the graph. Therefore, by the remark of Example 1, there are at least two red triangles T_1, T_4 in the graph. Since the shortest side of T_1 is red, it is the longest side of some triangle T_3. Similarly, the shortest side of T_4 is the longest side of some triangle T_2. This completes the proof.

7. Colour the squares of each face of $9 \times 9 \times 9$ cubes in black and white alternately, such that the squares on the corners are in black. Then

   1. Each face has 41 black squares and 40 white squares;
   2. A 2×1 grid is in one colour if and only if it is a bent one.

   Thus, among all the 2×1 bent grids, the number of black squares is 3 more than the white squares. Therefore, the number of bent grids is odd.

8. The answer is negative. Colour 24 squares of the chessboard as shown in Fig. 14(Ex).4. It is easy to write numbers on the coloured squares such that their sum S is odd. Since each 2×2, 3×3, or 4×4 sub-chessboard

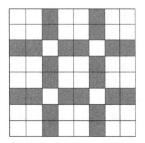

Figure 14(Ex).4

covers an even number of these coloured squares, S increases by an even number after an operation. So, S is always odd.

9. Let the set of small 12 squares in the grid occupied by the dominoes be C. Denote the small square in the upper right corner and the small square in the down left corner by B and D, respectively. Denote the set of squares on the upper left of $B \cup C \cup D$ by A; denote the set of squares to the lower right of $B \cup C \cup D$ by E.

Colour the grid in black and white alternately such that B is white. Then D is black; A contains 16 white squares and 13 black squares; E contains 16 black squares and 13 white squares.

We see that any domino can only be placed either in $A \cup B \cup D$ or $E \cup B \cup D$, that each contains at most 14 dominoes, since each domino occupies one black and one white square. So, we can put at most 28 dominoes on the grid. Fig. 14(Ex).5 is an example of 28 dominoes. So, the answer is 28.

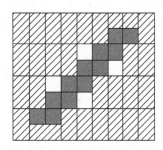

Figure 14(Ex).5

Chapter 15

1. Assign a cup placed bottom-down with number 1 and a cup placed bottom-up with number -1. Then calculate the product of these 11 numbers after each operation.

   At the beginning, 11 cups are all bottom-down, the product is 1. When turning over an even number of cups, the product remains 1. So, 11 cups cannot be all bottom-up (with the product -1) by any operation.

2. Assign each side with number 0 if its two endpoints are in the same colour, with number 1 otherwise. Then assign each small triangle with the sum of the numbers of the three sides. Let the sum of all the assigned numbers of the small triangles be S. S is odd, since each side, except sides AB, BC, and CA, is counted twice. On the other hand, $S = 3x + 2y$, where x is the number of small triangles whose vertices are of three colours and y is the number of small triangles with two vertices in the same colour. Thus, x is odd.

3. Assign each vertex of small square $P_i (i = 1, 2, \ldots, (n+1)^2)$ with a number a_i: if P_i is red, $a_i = 1$; if P_i is blue, $a_i = -1$.

   Label n^2 small squares with $j = 1, 2, \ldots, n^2$, and assign the jth small square a number A_j, A_j equals the product of the a_i's at its four vertices: if the square has just three vertices in a colour, $A_j = -1$, otherwise $A_j = 1$. So, the problem is equivalent to proving the number of -1 in $A_1, A_2, \ldots, A_{n^2}$ is even.

   Now considering the product $A_1 A_2 \ldots A_{n^2}$. For an intersection point inside the big square, the number on the point appears four times in the product. For a point on the side but not a vertex of the big square, the number appears twice in the product, and on points A, B, C, D, each number appears once. Hence, the product is

$$A_1 A_2 \ldots A_{n^2} = 1 \times (-1) \times 1 \times (-1) = 1.$$

Therefore, the number of -1 in $A_1, A_2, \ldots, A_{n^2}$ is even. That is, the number of small squares that each has just three vertices in a colour is even.

4. The answer is negative. If there is a stone on the square (i,j), then assign the square with number λ^{i+j}, $i, j \in \mathbb{N}^+$, where $\lambda = (\sqrt{5} - 1)/2$, otherwise with number 0.

   If a stone is at square (i,j), and there are no stones at $(i,j+1)$ and $(i,j+2)$, then at the next step, remove the stone at (i,j), and put stones at $(i,j+1)$ and $(i,j+2)$. After this move, the sum of the numbers on the board changes by

$$(\lambda^{i+(j+1)} + \lambda^{i+(j+2)}) - \lambda^{i+j} = \lambda^{i+j}(\lambda + \lambda^2 - 1) = 0,$$

   that is, the sum S of the numbers on the board is invariant. It is also invariant if stones are put at $(i+1,j)$ and $(i+2,j)$. At the beginning, $S = \lambda^2$.

   Suppose, on the contrary, after some finite steps, all (finite) stones are on the right of the fourth column. Then

$$S < \sum_{i=1}^{\infty} \sum_{j=5}^{\infty} \lambda^{i+j} = \lambda^6 \left(\sum_{i=0}^{\infty} \lambda^i \right)^2 = \lambda^6 (1-\lambda)^{-2} = \lambda^6 \lambda^{-4} = \lambda^2.$$

   A contradiction.

5. The answer is negative.

   Suppose that the starting position of B is the right neighbour of A.

   Assign each grid one of three numbers $1, \omega$, and $\omega^2 (\omega = (-1+i\sqrt{3})/2)$, alternately, such that the number of a grid is ω times the number of its left neighbouring grid, and is ω^2 times the number of its down-left neighbouring grid. By the rule of the game, any round jumps do not change the ratio of A and B. If, at the beginning, A is at a grid with number 1, then B is at a grid with number ω. If at last A and B can interchange their positions, then, $\frac{1}{\omega} = \frac{\omega}{1}$, a contradiction.

6. First, fill in the squares of the 3000×3000 grid at position (i,j) for a number $a_{ij} \equiv (i+j-1) \pmod 3 \in \{1,2,3\}$ as shown in Fig. 15(Ex).1. Denote the three colours by 1, 2, and 3.

   Colour a domino in colour k, if it does not cover number k.

   Consider a domino A, and suppose it is horizontally placed with colour 3, so it covers numbers 1 and 2 as shown in Fig. 15(Ex).2. There are

1	2	3	1	⋯
2	3	1	2	⋯
3	1	2	3	⋯
1	2	3	1	⋯
⋯	⋯	⋯	⋯	⋯

Figure 15(Ex).1

2	3	1	2
3	1	2	3
1	2	3	1

Figure 15(Ex).2

four squares with number 3 in the six neighbouring squares. Thus, there are at most 2 dominoes with colour 3 as A's neighbours.

Finally, we show that S_i ($1 \le I \le 3$), the number of dominoes in clour i, is 1,500,000. Obviously, $S_1 + S_2 + S_3 = 4,500,000$. Each square with number 1 is covered by a domino with colour 2 or 3, thus $S_2 + S_3 = 300,000$ and $S_1 = 1,500,000$. Similarly, $S_2 = S_3 = 1,500,000$. The proof is complete.

Chapter 16

1. Suppose, on the contrary, there were such a sequence of 17 terms $a_1, a_2, \ldots, a_{17}$ with positive sum of each consecutive 7 terms and negative sum of each consecutive 11 terms. Consider the following 7×11 matrix,

$$
\begin{pmatrix}
a_1 & a_2 & a_3 & \cdots & a_{11} \\
a_2 & a_3 & a_4 & \cdots & a_{12} \\
\cdots & \cdots & \cdots & \cdots & \cdots \\
a_7 & a_8 & a_9 & \cdots & a_{17}
\end{pmatrix}.
$$

The sum of each row is negative, and so is the sum of all entries of the matrix. On the other hand, the sum of each column is positive, and so is the sum of all entries of the matrix. A contradiction.

2. Count the number S of integral points in $A = \{(x,y) | 1 \le x \le n, 1 \le xy \le n\}$ in two ways.

On the one hand, for each fixed $i(i = 1, 2, \ldots, n)$, the number of integers y such that $1 \le iy \le n$, is $\lfloor n/i \rfloor$. So, the number of integral points in A on the line $x = i(i = 1, 2, \ldots, n)$ is $\lfloor n/i \rfloor$. Thus,

$$
S = \sum_{i=1}^{n} \left\lfloor \frac{n}{i} \right\rfloor. \tag{1}
$$

On the other hand, for each fixed $j(j = 1, 2, \ldots, n)$, the number of integral points such that $xy = j$ and $1 \le x \le n$ is just the number of positive factors $d(j)$ of j. Hence,

$$
S = \sum_{j=1}^{n} d(j). \tag{2}
$$

The proof is completed by (1) and (2).

3. If a teacher $t_i (1 \leq i \leq b)$ teaches two students $s_j, s_l (1 \leq j < l \leq c)$, we call the triple (t_i, s_j, s_l) a *teaching group*.

    On the one hand, by condition (a), for any i $(1 \leq i \leq b)$, the number of *teaching groups* containing teacher t_i is $\binom{k}{2}$. Hence, there are $b\binom{k}{2}$ *teaching groups*.

    On the other hand, by condition (b), for any $j, l (1 \leq j < l \leq c)$, the number of *teaching groups* containing students s_j, s_l is h, and there are $\binom{c}{2}$ selection ways for s_j, s_l. Hence, there are $h\binom{c}{2}$ *teaching groups*.

    Summing up, we have $b\binom{k}{2} = h\binom{c}{2}$, that is, $\frac{b}{h} = \frac{c(c-1)}{k(k-1)}$.

4. Suppose, on the contrary, for any two persons A and B, at least 6 of the other 10 people know only one of A and B. Now we count the number N of triples (A, B, C), where C knows only one of A and B.

    On the one hand, there are $\binom{12}{2} = 66$ different pairs of (A, B), and for each pair, there are at least 6 choices of C. Hence, $N \geq 66 \times 6 = 396$.

    On the other hand, fix C, which has 12 choices. If C knows n persons, then the number of pairs of (A, B) is

$$n(11 - n) \leq 30.$$

    Thus, $N \leq 12 \times 30 = 360$. A contradiction.

5. For $1 \leq i \leq m, 1 \leq j \leq n$, denote the entry of the matrix at (i, j) by a_{ij}, and the sum of the i-th row by r_i, the sum of the jth column by c_j. Then, $r_i, c_j > 0$. If $a_{ij} > 0$, then $r_i = c_j$. Thus, we always have $\frac{a_{ij}}{r_i} = \frac{a_{ij}}{c_j}$.

    We count $S = \sum_{i=1}^{m} \sum_{j=1}^{n} \frac{a_{ij}}{r_i}$ in two ways.
    On the one hand,

$$S = \sum_{i=1}^{m} \frac{1}{r_i} \cdot \sum_{j=1}^{n} a_{ij} = \sum_{i=1}^{m} \frac{1}{r_i} \cdot r_i = m.$$

On the other hand,

$$S = \sum_{j=1}^{n} \sum_{i=1}^{m} \frac{a_{ij}}{c_j} = \sum_{j=1}^{n} \frac{1}{c_j} \cdot \sum_{i=1}^{m} a_{ij} = \sum_{j=1}^{n} \frac{1}{c_j} \cdot c_j = n.$$

So, $m = n$.

6. Denote $B = (b_{ij})_{1 \leq i,j \leq n}$. Define

$$S = \sum_{1 \leq i,j \leq n} (x_i + y_j)(a_{ij} - b_{ij}).$$

On the one hand,

$$S = \sum_{i=1}^{n} x_i \left(\sum_{j=1}^{n} a_{ij} - \sum_{j=1}^{n} b_{ij} \right) + \sum_{j=1}^{n} y_j \left(\sum_{i=1}^{n} a_{ij} - \sum_{i=1}^{n} b_{ij} \right) = 0.$$

On the other hand, for any $i, j (1 \leq i, j \leq n)$, we have $(x_i + y_j)(a_{ij} - b_{ij}) \geq 0$. In fact, if $a_{ij} = 1$, $x_i + y_j \geq 0$, then $a_{ij} - b_{ij} = 1 - b_{ij} \geq 0$; if $x_i + y_j < 0$, then $a_{ij} - b_{ij} = -b_{ij} \leq 0$.

So, all equalities hold, that is, $(x_i + y_j)(a_{ij} - b_{ij}) = 0$ holds for any i, j.

Especially, if $a_{ij} = 0$, then $x_i + y_j < 0$, so $a_{ij} - b_{ij} = 0$, that is $b_{ij} = 0$. So, for all i, j, we have $a_{ij} \geq b_{ij}$. But the sum of elements of any row of B is equal to the sum of elements of the corresponding row of A, so $a_{ij} = b_{ij}$ for all i, j. That is, $B = A$.

Chapter 17

1. Fix $x_1, x_2, \ldots, x_{n-2}$, and note that
$$(1 - x_{n-1})(1 - x_n) = 1 - (x_{n-1} + x_n) + x_{n-1}x_n \geq 1 - (x_{n-1} + x_n),$$
So,
$$(1 - x_1)(1 - x_2) \cdots (1 - x_n) \geq (1 - x_1)(1 - x_2) \cdots (1 - x_{n-2})$$
$$(1 - (x_{n-1} + x_n)).$$
Repeat the similar process $n - 2$ times, and we see that
$$(1 - x_1)(1 - x_2) \cdots (1 - x_n) \geq 1 - (x_1 + x_2 + \cdots + x_n) \geq 1/2.$$
The equality holds if $x_1 = \frac{1}{2}, x_2 = x_3 = \cdots = x_n = 0$, thus, the minimum is $1/2$.

2. There are only a finite number of ways to write 49 as the sum of 10 positive integers $x_1, x_2, \ldots, x_{10}$. So, $S = x_1^2 + x_2^2 + \cdots + x_{10}^2$ has extremes.

Lemma 1. *Let $x_1 \leq x_2 \leq \cdots \leq x_n$, $m \geq n$ be positive integers, such that $x_1 + x_2 + \cdots + x_n = m$, $S = x_1^2 + x_2^2 + \cdots + x_n^2$.*
Then S attains its maximum at $x_1 = x_2 = \cdots = x_{n-1} = 1$, $x_n = 1 + m - n$.

Proof. Suppose, on the contrary, $x_j \geq 2$, $1 \leq j \leq n - 1$. So,
$$(x_j - 1) + (x_n + 1) = x_j + x_n.$$
Replace x_j and x_n by $x_j - 1$ and $x_n + 1$, and we have
$$((x_j - 1)^2 + (x_n + 1)^2) - (x_j^2 + x_n^2) = 2(x_n - x_j) + 2 \geq 2.$$
A contradiction.

265

Therefore, the maximum of $S = x_1^2 + x_2^2 + \cdots + x_{10}^2$ is $9 + 40 \times 40 = 1609$.

Lemma 2. $S = x_1^2 + x_2^2 + \cdots + x_n^2$ attains its minimum at

$$x_1 = x_2 = \cdots = x_t = \lfloor m/n \rfloor,$$

$$x_{t+1} = x_{t+2} = \cdots = x_n = \lfloor m/n \rfloor + 1,$$

where $t = n - (m - n\lfloor m/n \rfloor)$.

Proof. Suppose, on the contrary, there are some $x_j \leq \lfloor m/n \rfloor - 1$ and $x_i \geq \lfloor m/n \rfloor + 1$. Replace x_j and x_i by $x_j + 1$ and $x_i - 1$, and we have

$$(x_j + 1)^2 + (x_i - 1)^2 - (x_j^2 + x_i^2) = -2(x_i - x_j - 1) \leq -2.$$

A contradiction.

Therefore, $x_1^2 + x_2^2 + \cdots + x_{10}^2$ attains its minimum 241 at $x_1 = 4, x_2 = x_3 = \cdots = x_{10} = 5$.

3. Since there are only a finite number of ways to write 2006 as the sum of distinct positive integers, there must be a way such that the product of these positive integers attains its maximum.

Let the distinct positive integers be $x_1 < x_2 < \cdots < x_n$ such that $x_1 + x_2 + \cdots + x_n = k$, $k \geq (n+1)(n+1)/2 - 1$, and $P = x_1 x_2 \cdots x_n$ attains its maximum $P_{\max}$. First, we see that $x_1 \geq 2$, otherwise the product $x_2 x_3 \ldots x_{n-1}(x_n + 1)$ will be larger. Next, we see that if $a, b \in \mathbb{N}^+$, $a \geq 2$, then $a(a + b) > 2a + b$. So, the more the factors, the larger the product. Let $p \geq n$ be the smallest positive integer, such that

$$2 + 3 + \cdots + p = p(p + 1)/2 - 1 \geq k, \tag{1}$$

that is, $p = \lceil (\sqrt{8k + 9} - 1)/2 \rceil$.

Then, let $m = p(p+1)/2 - 1 - k$. We see that $m < p$ by the definition of p.

If $m = 0$, then $n = p - 1$, $x_i = i + 1$, $i = 1, 2, \ldots, n$. $P_{\max} = p!$.

If $m = 1$, then in (1), remove 2, and replace p with $p + 1$, that is, $n = p - 2$, $x_i = i + 2$, $i = 1, 2, \ldots, n - 1$, $x_n = p + 1$. $P_{\max} = (p + 1)!/2$.

If $m \in \{2, 3, \ldots, p - 1\}$, then in (1), just remove m, that is, $n = p - 2$, $x_i = i + 1$, for $i = 1, 2, \ldots, m - 2$; $x_i = i + 2$, for $i = m - 1, m, \ldots, n$. $P_{\max} = p!/m$.

For $k = 2006$, $p = 63$, $m = 9$, $n = 61$. So,

$$2006 = (2 + 3 + \cdots + 8) + (10 + 11 + \cdots + 63),$$

the largest product is $\frac{63!}{9}$.

4. For positive integer m, define $f(m)$ as the number of primes in n consecutive positive integers $m, m+1, \ldots, m+n-1$. Obviously, $f(1) = \pi(n)$.

Take consecutive positive integers $(n+1)! + 2, (n+1)! + 3, \ldots, (n+1)! + (n+1)$, which are all composite. So, $f((n+1)! + 2) = 0$. Hence,

$$f((n+1)! + 2) \leq k \leq \pi(n) = f(1). \tag{1}$$

On the other hand, consider the relation between $f(m+1)$ and $f(m)$.

If m and $m+n$ are both prime or composite, we have $f(m+1) = f(m)$;
If m is prime and $m+n$ is composite, we have $f(m+1) = f(m) - 1$;
If m is composite and $m+n$ is prime, we have $f(m+1) = f(m) + 1$.
That is, $|f(m+1) - f(m)| \leq 1$ for all $m \in \mathbb{N}^+$, associated with (1), we see that there exists some $i \in \{1, 2, \ldots, (n+1)! + 2\}$ such that $f(i) = k$, that is, in consecutive n positive integers $i, i+1, \ldots, i+n-1$ there exist(s) exactly k prime(s).

5. **Lemma.** *Suppose that $x_1 \leq x_2 \leq \cdots \leq x_n$ are integers, such that $x_1 + x_2 + \cdots + x_n = m$, where m, n are positive integers. Then $S = \sum_{i-1}^{n} x_i^2 - \frac{m^2}{n}$ attains its minimum if and only if $0 \leq x_n - x_1 \leq 1$.*

Proof. If $x_n \geq x_1 + 2$, then there exist $n \geq j > i \geq 1$ such that $x_1 = x_i < x_{i+1}$ and $x_{j-1} < x_j = x_n$. So,

$$(x_i + 1)^2 + (x_j - 1)^2 = x_i^2 + x_j^2 - 2(x_j - x_i - 1) < x_i^2 + x_j^2.$$

That is, we can get a smaller S by increasing x_i by 1 and decreasing x_j by 1. Such a procedure can continue until $x_n - x_1 \leq 1$ and S attains its minimum.

Remark. By the lemma, we can see that if S attains its minimum at $x_1 \leq x_2 \leq \cdots \leq x_n$, then $x_1 \geq 0$. In fact, since $m \geq 1$, if $x_1 \leq -1$, then $x_n \geq 1$, and $x_n - x_1 \geq 2$, contradicting the lemma.

In the following, we solve (a) and (b) simultaneously.

Let $m = nk - t$, $k \in \mathbb{N}^+$, $t \in \{0, 1, \ldots, n-1\}$.

Let $x_1 = x_2 = \cdots = x_t = k - 1, x_{t+1} = x_{t+2} = \cdots = x_n = k$ with sum m. By the lemma, S attains its minimum.

$$S = \sum_{i=1}^{n} x_i^2 - \frac{m^2}{n}$$

$$= \sum_{i=1}^{n} x_i^2 - \frac{1}{n} \sum_{i=1}^{n} x_i^2 - \frac{2}{n} \sum_{1 \le i < j \le n} x_i x_j$$

$$= \frac{1}{n} \sum_{1 \le i < j \le n} (x_j - x_i)^2 = \frac{t(n-t)}{n} < 2.$$

Thus, $f(t) = t^2 - nt + 2n = (t - n/2)^2 + n(8 - n)/4 > 0$ holds for all t if $n \le 7$. That is, when $n \le 7$, there exist integer solutions for all positive integers m.

If $n \ge 8$, $f(4) = 16 - 2n \le 0$. So, for $m = nk - 4$ and $n \ge 8$, there are no solutions.

Thus, the answer of (a) is $n = 1, 2, \ldots, 7$.

Now, we find all the pairs of (m, n) for $n \ge 8$ such that there are integer solutions.

By computation, $f(0) = 2n > 0$, $f(1) = f(n - 1) = n + 1 > 0$, $f(2) = f(n - 2) = 4 > 0$. And if $n = 8$, $f(3) = f(5) = 1 > 0$. So, we can take $t = 0, 1, 2, 3, 5, 6, 7$ for $n = 8$.

If $n \ge 9$, $f(3) = f(n - 3) = 9 - n \le 0$. By the property of quadratic functions, if $t \in (3, n - 3)$, $f(t) < \max\{f(3), f(n - 3)\} \le 0$. So, we can just take $t = 0, 1, 2, n - 2, n - 1$. Combining that $m = nk - t, k = 1, 2, \ldots, t \in \{0, 1, \ldots, n - 1\}$, we derive the answer to (b) as: all (m, n) are classified as follows:

1. $(a, 1), (a, 2), \ldots, (a, 7)$, where a is any positive integer;
2. $(bk - t, b)$, where integer $b \ge 8$ and k is any positive integer, $t = 0, 1, 2, b - 2, b - 1$,
3. $(8k - 3, 8), (8k - 5, 8)$, where k is any positive integer.

And all the solutions have the form

$$x_1 = x_2 = \cdots = x_t = k - 1, x_{t+1} = x_{t+2} = \cdots = x_n = k.$$

6. Let the number of points in each group of the 30 groups be $n_1, n_2, \ldots, n_{30}$, respectively. The problem is: under the conditions of $n_1 < n_2 < \cdots < n_{30}$, and $\sum_{i=1}^{30} n_i = 1989$, find the maximum of the

number of triangles

$$S = \sum_{1 \le i < j < k \le 30} n_i n_j n_k.$$

Since partitions of 1989 points into 30 groups are finite, the maximum of S exists. If S attains its maximum, then

(1) $n_{i+1} - n_i \le 2(i = 1, 2, \ldots, 29)$. Suppose, on the contrary, say, $n_2 - n_1 > 2$, then,

$$S = n_1 n_2 \sum_{k=3}^{30} n_k + (n_1 + n_2) \sum_{3 \le j < k \le 30} n_j n_k + \sum_{3 \le i < j < k \le 30} n_i n_j n_k.$$

Let $n_1' = n_1 + 1$, $n_2' = n_2 - 1$. Then $n_3 > n_2' > n_1' > n_1$, and

$$n_1' n_2' = (n_1 + 1)(n_2 - 1) = n_1 n_2 + (n_2 - n_1 - 1) > n_1 n_2.$$

Replace n_1, n_2 with n_1', n_2', the first term $n_1 n_2 \sum_{k=3}^{30} n_k$ of S increases, and the other terms remain unchanged. A contradiction.

(2) There is at most one i such that $n_{i+1} - n_i = 2$. Otherwise, there are $j > i$, $n_{i+1} - n_i = 2$, $n_{j+1} - n_j = 2$. Then let $n_i' = n_i + 1$, $n_j' = n_j - 1$, and S will increase.

(3) There exists i_0, such that $n_{i_0+1} - n_{i_0} = 2$. Since, otherwise,

$$n_1 + n_2 + \cdots + n_{30} = 15 \times (2n_1 + 29) = 1989.$$

But 1989 is not divisible by 15.
So, let $n_1, n_2, \ldots, n_{30}$ be as follows:

$$n_1, n_1 + 1, \ldots, n_1 + i_0 - 1, n_1 + i_0 + 1, \ldots, n_1 + 30, 1 \le i_0 \le 29,$$

satisfying $1989 = (2n_1 + 30) \times 31 - (n_1 + i_0)$. That is,

$$30n_1 - i_0 = 1524.$$

The solution is $n_1 = 51$, $i_0 = 6$. Thus, S attains its maximum at $(n_1, n_2, \ldots, n_{30}) = (51, 52, \ldots, 56, 58, 59, \ldots, 81)$.

7. Assume $a \le b \le c \le d$, and denote

$$f(a, b, c, d) = \frac{1}{a} + \frac{1}{b} + \frac{1}{c} + \frac{1}{d} + \frac{k}{a + b + c + d}.$$

We shall prove a more general inequality

$$\frac{1}{a} + \frac{1}{b} + \frac{1}{c} + \frac{1}{d} + \frac{k}{a+b+c+d} \geq 4 + \frac{k}{4}, \quad \text{for} \quad 0 \leq k \leq 12, \quad (1)$$

in which $k = 9$ is the desired inequality.

First, we show that

$$f(a, b, c, d) \geq f(\sqrt{ac}, b, \sqrt{ac}, d). \quad (2)$$

In fact, (2) is equivalent to

$$\frac{1}{a} + \frac{1}{c} + \frac{k}{a+b+c+d} \geq \frac{1}{\sqrt{ac}} + \frac{1}{\sqrt{ac}} + \frac{k}{2\sqrt{ac}+b+d}$$

$$\Leftrightarrow \frac{(\sqrt{a}-\sqrt{c})^2}{ac} \geq \frac{k(\sqrt{a}-\sqrt{c})^2}{(a+b+c+d)(2\sqrt{ac}+b+d)}$$

$$\Leftarrow (a+b+c+d)(2\sqrt{ac}+b+d)$$

$$\geq kac, \left(\text{notice that } b+d \geq 2\sqrt{bd} = \frac{2}{\sqrt{ac}}\right)$$

$$\Leftarrow \left(a+c+\frac{2}{\sqrt{ac}}\right)\left(2\sqrt{ac}+\frac{2}{\sqrt{ac}}\right) \geq kac. \quad (3)$$

Since $1 = abcd \geq a \cdot a \cdot c \cdot c \Rightarrow ac \leq 1 \Rightarrow \frac{2}{\sqrt{ac}} \geq 2\sqrt{ac}$, and $a+c \geq 2\sqrt{ac}$, it follows that, the left-hand side of (3) $\geq (2\sqrt{ac}+\frac{2}{\sqrt{ac}})(2\sqrt{ac}+\frac{2}{\sqrt{ac}}) \geq 4\sqrt{ac} \cdot 4\sqrt{ac} = 16ac > kac$.

Inequality (2) shows that the minimum of $f(a, b, c, d)$ is attained at $a = c$, that is, $a = b = c \leq 1$. So, let $(a, b, c, d) = (t, t, t, t^{-3})$, and, it suffices to show, $\forall t \in (0, 1]$, we always have,

$$f(t, t, t, t^{-3}) \geq 4 + \frac{k}{4}. \quad (4)$$

Obviously, $f(1, 1, 1, 1) = 4 + k/4$. For $0 < t < 1$, we see that

$$f(t, t, t, t^{-3}) \geq 4 + \frac{k}{4} \Leftrightarrow \frac{3}{t} + t^3 + \frac{k}{3t+t^{-3}} \geq 4 + \frac{k}{4}$$

$$\Leftrightarrow (3t^{-1} + t^3 - 4) \geq k\left(\frac{1}{4} - \frac{1}{3t+t^{-3}}\right)$$

$$\Leftarrow 12 \leq \frac{3t^{-1} + t^3 - 4}{\frac{1}{4} - \frac{1}{3t+t^{-3}}} = \frac{4(3 - 4t + t^4)(1 + 3t^4)}{(1 - 4t^3 + 3t^4)t}$$

$$= \frac{4(3 + 2t + t^2)(1 + 3t^4)}{(1 + 2t + 3t^2)t}, \tag{5}$$

$$\Leftarrow 3t^6 + 6t^5 + 9t^4 - 9t^3 - 5t^2 + 3 \geq t \text{ (since } 6t^5 \geq 6t^6, -9t^3 \geq -9t^2)$$

$$\Leftarrow g(t) = 9t^6 + 9t^4 - 14t^2 + 3 \geq t$$

$$\Leftrightarrow g(x) = 9x^3 + 9x^2 - 14x + 3 \geq \sqrt{x}, \quad \text{for} \quad x \in (0,1). \tag{6}$$

The proof of (6). Since $g'(x) = 27x^2 + 18x - 14$, $g''(x) = 54x + 18 > 0$, the minimum of $g(x)$ is attained at $x = x_0 = \frac{\sqrt{51}-3}{9}$, which is the positive root of $g'(x) = 0$. By

$$g(x) = (27x^2 + 18x - 14)\left(\frac{1}{3}x + \frac{1}{9}\right) + 3 + \frac{14 - 66x}{9},$$

so, $g(x) \geq g(x_0) = 3 + \frac{14-66x_0}{9} \approx 3 - 1.81894 > 1.18 > \sqrt{x}$.

Remark. The k in (1) has a best possible upper bound $k_0 = 12.4185242939873\ldots$, which is the minimum of the right-hand side of (5) subject to the condition $t \in [0,1]$. The minimum attains at $t = 0.692466380367\ldots$, which is the real root of $9t^5 + 11t^4 + 6t^3 - 6t^2 - 3t - 1 = 0$. Note that, we can prove that $k_0 = 16x$, where x is the real root of

$$729x^5 + 3923x^4 + 9002x^3 - 42x^2 - 5107x - 1849 = 0.$$

It is a common fact that the extremum of a rational function is a zero of a polynomial, when the extremum point is an inner point, and the polynomial is of rational coefficients if the rational function is.

Chapter 18

1. (1) $(2k^2)^2 + (2k)^2 + 1 = (2k^2 + 1)^2$;

   (2) $2(k^2 + k)^2 + 2(k^2 - k)^2 + 1 = (2k^2 + 1)^2$;

   (3) $(k^2)^2 + ((k+1)^2)^2 + 1 = 2(k^2 + k + 1)^2$.

2. Construct a function $f(x, y, z) = (x^3 - x) + (y^3 - y) + (z^3 - z)$, then the first equation of the system is equivalent to $f(x, y, z) = 0$.

   If $x, y, z \geq 1$, then the equality of $f(x, y, z) \geq 0$ holds if and only if $x = y = z = 1$.

   But the second equation of the system does not hold for $x = y = z = 1$. So, if any solution exists, at least one unknown is less than 1, say $x < 1$, then $x^2 + y^2 + z^2 > y^2 + z^2 > 2yz > xyz$, a contradiction. So, no positive solution exists.

3. Draw a figure as shown in Fig. 18(Ex).1, point A being the origin, $AB = 12$, $AC \perp AB$, $BD \perp AB$, $AC = 1$, $BD = 4$. Let point P be on the x-axis such that $AP = x$, then $f(x) = |CP| + |DP|$ attains its minimum when point P is on the line CD, so $f_{\min} = |CD| = |AE| = \sqrt{12^2 + 5^2} = 13$.

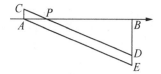

Figure 18(Ex).1

4. The system $\begin{cases} x^3 + \sin x - 2a = 0, \\ 4y^3 + \frac{1}{2}\sin 2y + a = 0 \end{cases}$ yields $\begin{cases} x^3 + \sin x = 2a, \\ 8y^3 + \sin 2y = -2a. \end{cases}$

273

Construct an increasing function $f(t) = t^3 + \sin t$. Since $f(x) = -f(2y) = f(-2y)$, we see that $x + 2y = 0$. So, $\cos(x + 2y) = 1$.

5. From $x^3 - 3x^2 + 5x - 3 = (x-1)^3 + 2(x-1)$, construct an increasing odd function $f(t) = t^3 + 2t$.

Since $\begin{cases} \alpha^3 - 3\alpha^2 + 5\alpha = 1, \\ \beta^3 - 3\beta^2 + 5\beta = 5. \end{cases} \Rightarrow \begin{cases} \alpha^3 - 3\alpha^2 + 5\alpha - 3 = -2, \\ \beta^3 - 3\beta^2 + 5\beta - 3 = 2. \end{cases}$

We see that $f(\alpha - 1) = -f(\beta - 1) = f(1 - \beta)$.

Since $f(x)$ is increasing, $\alpha - 1 = 1 - \beta$. So, $\alpha + \beta = 2$.

6. If $n = 1$, the solution is $x_1 = 1$.

Assume $n \geq 2$. Construct a function

$$f(t) = (t-x_1)(t-x_2)\cdots(t-x_n) = t^n + a_1 t^{n-1} + a_2 t^{n-2} + \cdots + a_{n-1}t + a_n.$$

Then, $f(x_1) = f(x_2) = \cdots = f(x_n) = 0$. On the other hand, with the given equations, it follows that

$$0 = \sum_{i=1}^{n} f(x_i) = \sum_{i=1}^{n} x_i^n + a_1 \sum_{i=1}^{n} x_i^{n-1} + a_2 \sum_{i=1}^{n} x_i^{n-2}$$

$$+ \cdots + a_{n-1} \sum_{i=1}^{n} x_i + a_n \sum_{i=1}^{n} 1$$

$$= n + a_1 n + a_2 n + \cdots + a_{n-1}n + a_n n = nf(1).$$

So, $f(1) = 0$, which implies there is a 1 in $x_1, x_2, \ldots, x_n$, say, $x_n = 1$. Then the rest unknowns $x_1, x_2, \ldots, x_{n-1}$ satisfy $\sum_{i=1}^{n-1} x_i^k = n - 1(k = 1, 2, \ldots, n)$. In the similar way, $x_i = 1$ for all $i = 1, 2, \ldots, n - 1$. Thus, the solution is $x_1 = x_2 = \cdots = x_n = 1$.

7. (a) The answer is negative, since $1 + 2 + \cdots + 96 = \frac{96 \times (96+1)}{2} = 48 \times 97$ is not divisible by 32.

(b) The answer is positive. The sum of the elements of each triple set is $\frac{1+2+\cdots+99}{33} = \frac{99 \times (99+1)}{33 \times 2} = 150$. First, partition $1, 2, 3, \ldots, 66$ into 33 pairs such that the sums of the paired elements consist of an arithmetic sequence with common difference 1 as follows:

$$1 + 50, 3 + 49, \ldots, 33 + 34, 2 + 66, 4 + 65, \ldots, 32 + 51.$$

Then, form the 33 triple sets as follows, such that each triple set has the same sum:

$$\{1, 50, 99\}, \{3, 49, 98\}, \ldots, \{33, 34, 83\}, \{2, 66, 82\}, \{4, 65, 81\}, \ldots,$$
$$\{32, 51, 67\}.$$

Remark. The problem can be generalized to the following:

Find all positive integers such that the triple subsets $A_i = \{x_i, y_i, z_i\}$, $i = 1, 2, \ldots n$, of $M = \{1, 2, 3, \ldots, 3n\}$, satisfy $A_1 \cup A_2 \cup \cdots \cup A_n = M$, for any $i, j (1 \leq i \neq j \leq n)$, $s_i = s_j$, where $s_i = x_i + y_i + z_i$.

Solution. First, we require that $n | (1 + 2 + 3 + \cdots + 3n)$, that is,

$$n \left| \frac{3n(3n+1)}{2} \right. \Rightarrow 2 | 3n + 1.$$

So, n is odd.

If n is odd, divide $1, 2, 3, \ldots, 2n$ into n pairs, such that the sums of the paired elements consist of an arithmetic sequence with common difference 1 as follows:

$$1 + \left(n + \frac{n+1}{2}\right), 3 + \left(n + \frac{n-1}{2}\right), \cdots, n + (n+1);$$

$$2 + 2n, 4 + (2n - 1), \cdots, (n-1) + \left(n + \frac{n+3}{2}\right).$$

The sums are

$$x_k + y_k = \begin{cases} 2k - 1 + \left(n + \dfrac{n+1}{2} + 1 - k\right), & 1 \leq k \leq \dfrac{n+1}{2}, \\ [1 - n + 2(k-1)] + \left[2n + \dfrac{n+1}{2} - (k-1)\right], & \dfrac{n+3}{2} \leq k \leq n. \end{cases}$$

Since $x_k + y_k + 3n + 1 - k = \frac{9n+3}{2}$ is constant, let $z_k = 3n + 1 - k$ and we obtain n triple sets each with sum $(9n+3)/2$:

$$\left\{1, n + \frac{n+1}{2}, 3n\right\}, \left\{3, n + \frac{n-1}{2}, 3n - 1\right\}, \cdots, \left\{n, n+1, 3n + 1 - \frac{n+1}{2}\right\};$$

$$\left\{2, 2n, 3n + 1 - \frac{n+3}{2}\right\}, \cdots, \left\{n - 1, n + \frac{n+3}{2}, 2n + 1\right\}.$$

Then take the above sets as $A_1, A_2, \ldots, A_n$, respectively.

8. The answer is positive.

Take 2008 primes $p_1 < p_2 < \cdots < p_{2008}$. Construct the subsets $A_1, A_2, \ldots, A_{2009}$ of $\mathbb{N}^+$ as follows: A_1 is the set of all positive integers divisible by p_1; A_2 is all positive integers divisible by p_2 but not by p_1; $\ldots$; A_{2008} is the set of all positive integers divisible by p_{2008} but not by $p_1, p_2, \ldots, p_{2007}$; A_{2009} is the set of all positive integers not divisible by $p_1, p_2, \ldots, p_{2008}$.

Then $A_1, A_2, \ldots, A_{2009}$ are mutually disjoint with union $\mathbb{N}^+$.

Now, for any $x \in A_m, y \in A_n, m < n$, $p_m|x$, so, $p_m|xy$; on the other hand, x, y are not divisible by $p_1, p_2, \ldots, p_{m-1}$, so, xy is not divisible by $p_1, p_2, \ldots, p_{m-1}$. Hence, $xy \in A_m$.

Denote the 2008 colours by $1, 2, \ldots, 2008$, and colour the numbers in set A_i with colour i, $i = 1, 2, \ldots, 2008$. The required conditions are satisfied.

9. First, take $m = 1$, then $\frac{n+1}{m} + \frac{m+1}{n} = n + 1 + \frac{2}{n}$ is an integer if and only if $n = 2$.

Suppose that there are positive integers m, n $(m < n)$, such that $\frac{n+1}{m} + \frac{m+1}{n} = t \in \mathbb{N}$. Then,

$$tn = \frac{n(n+1)}{m} + m + 1.$$

We have $\frac{n(n+1)}{m} = s \in \mathbb{N}$. Moreover, $s > n$, since $n > m$. Thus,

$$tn = \frac{n(n+1)}{m} + m + 1 = s + \frac{n(n+1)}{s} + 1.$$

So, $\frac{s+1}{n} + \frac{n+1}{s} = t$.

That is, if (m, n) is a pair of positive integers such that $\frac{n+1}{m} + \frac{m+1}{n}$ is an integer, so is (n, s), with $s = \frac{n(n+1)}{m}$ and $s > n > m$. Therefore, there are infinitely many such pairs, for example, $(1, 2), (2, 6)$, $(6, 21), (21, 77) \ldots$ (they all correspond to the same integer $t = 4$). This completes the proof.

10. Every two points in k points on the circle are the endpoints of a minor arc and a major arc. So, there are at most $k(k-1)$ arc lengths.

If $k(k-1) \geq 20$, then $k \geq 5$;

If $k(k-1) \geq 30$, then $k \geq 6$.

On the other hand, if $k = 5$, $P_{21} = 5$, $T_{21} = (1, 3, 10, 2, 5)$.

For $n = 31$, if $k = 6$, the partition can be any one as in Fig. 18(Ex).2. So, $P_{31} = 6$, $T_{31} = (1, 2, 7, 4, 12, 5)$, $(1, 2, 5, 4, 6, 13)$, $(1, 3, 2, 7, 8, 10)$, $(1, 3, 6, 2, 5, 14)$ or $(1, 7, 3, 2, 4, 14)$, and so on.

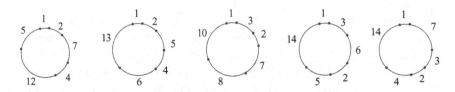

Figure 18(Ex).2

11. (a) Fix an integer $q > n$, and let $S_1 = \{n!q, n!q^2, \ldots, n!q^n\}$. Then, for any two non-empty subsets A, B of S_1, $f(A)$ and $f(B)$ are both positive integers. We show that if $A \neq B$, then $f(A) \neq f(B)$. Suppose, on the contrary, $f(A) = f(B)$, then

$$\sum_{n!q^i \in A} |B|q^i = \sum_{n!q^j \in B} |A|q^j. \tag{1}$$

Since $\max\{|A|, |B|\} \leq n < q$, we regard the positive integers on both sides of (1) as their base-q representations. So, $|A| = |B|$, and $i = j$. That is, $A = B$, a contradiction.

(b) By definition of S_2, we see that, for any two different non-empty subsets A, B of S_2, there are two non-empty subsets A_1, B_1 of S_1, satisfying $|A_1| = |A|$, $|B_1| = |B|$, and

$$f(A) = K!xf(A_1) + 1, \quad f(B) = K!xf(B_1) + 1. \tag{2}$$

We shall show that $f(A)$ and $f(B)$ are coprime. Suppose that $f(A)$ and $f(B)$ have a common factor $d > 1$, and so by (2), d is a factor of $f(A)f(B_1) - f(B)f(A_1) = f(B_1) - f(A_1)$. Since $|f(B_1) - f(A_1)|$ is a positive integer less than K, $d|K!$. Further, by (2) and $d|f(A)$, we see that $d = 1$, which completes the proof.

(c) We can choose $2^n - 1$ distinct primes $p_i > K$, $i = 1, 2, \ldots, 2^n - 1$. Denote the average of elements of every non-empty subset of S_1 by $\alpha_1, \alpha_2, \ldots, \alpha_{2^n-1}$. Then,

$$(p_i, K!\alpha_i) = 1, \quad (p_i^2, p_j^2) = 1, \quad i, j = 1, 2, \ldots, 2^n - 1, i \neq j.$$

Therefore, by the Chinese remainder theorem, the system of congruence equations

$$K!x\alpha_i \equiv -1 \pmod{p_i^2}, \quad i = 1, 2, \ldots, 2^n - 1$$

has a positive integer solution x. Then for any non-empty subset A of S_2 with this x, $f(A) = K!xf(A_1) + 1$ is a multiple of some p_i^2, and thus, a composite number.

Chapter 19

1. The answer is negative. Since

$$\left[\frac{\sqrt{2}}{2}(a+b)\right]^2 + \left[\frac{\sqrt{2}}{2}(a-b)\right]^2 = a^2 + b^2,$$

the sum of squares of the three numbers is invariant under the operation. At the beginning, $5^2 + 12^2 + 18^2 = 493$, while at the end, $3^2 + 13^2 + 20^2 = 578 > 493$. Thus, 5,12,18 cannot become 3,13,20.

2. Suppose there are $n > 1$ numbers on the blackboard. Erase two numbers, say a_{n-1} and a_n, with number $a'_{n-1} = a_{n-1} + a_n + a_{n-1}a_n$. Then

$$(1 + a_{n-1})(1 + a_n) = 1 + a_{n-1} + a_n + a_{n-1}a_n = 1 + a'_{n-1}.$$

So, $\prod_{i=1}^{n}(1 + a_i) = (1 + a'_{n-1})\prod_{i=1}^{n-2}(1 + a_i)$ is invariant under the operation. Suppose x is the last number that remains on the blackboard, then

$$x + 1 = \prod_{i=1}^{100}\left(1 + \frac{1}{i}\right) = \prod_{i=1}^{100}\frac{i+1}{i} = 101.$$

That is, $x = 100$.

3. The answer is negative, since the 4×4 grid as shown in Fig. 19(Ex).1 is invariant under the operation.

1	−1	−1	1
−1	1	1	−1
−1	1	1	−1
1	−1	−1	1

Figure 19(Ex).1

4. The answer is negative. We use a triple (a, b, c) to denote a chameleons in grey, b in brown, and c in purple. Then the triple (a, b, c) may change to $(a − 1, b − 1, c + 2)$ or $(a − 1, b + 2, c − 1)$ or $(a + 2, b − 1, c − 1)$ each time.

We see that $a − b$ may change to $a − b$ or $a − b − 3$ or $a − b + 3$ each time. That is, $(a − b) \pmod 3$ is an invariant. If one of a, b, and c is 45, and others are 0, then $a − b \equiv 0 \pmod 3$. But at the very beginning, $a − b \equiv 13 − 15 \equiv −2 \equiv 1 \pmod 3$.

5. The answer is negative. Denote the coordinates of the piece P_i by $P_i(x, y)$ $i = 1, 2, 3, 4$. If there were $P_1 = (0, 0)$, $P_2 = (1, 1)$, $P_3 = (3, 0)$, $P_4 = (2, −1)$ after some finite moves. We see that $x − y \equiv 0 \pmod 3$ for all $i = 1, 2, 3, 4$. And the coordinates of previous move of any P_i are $2P_j − P_i$ for some $j \in \{1, 2, 3, 4\}$. So $x − y \equiv 0 \pmod 3$ is invariant. But initially, P_2 at $(1,0)$, and $x − y \equiv 1 \pmod 3$.

6. Consider the length L of the boundary of black regions (include the inner boundary).

If one colours a square P that has $k (\geq 2)$ neighbouring black squares in black, then L changes to $L − k + (4 − k) = L + 4 − 2k \leq L$. So, L is monotonically decreasing.

At the beginning, colour any $n − 1$ squares in black, so $L \leq 4(n − 1)$.

If all squares are coloured in black, then the boundary has length $L = 4n > 4(n − 1)$. This proves the conclusion.

7. For a simple fraction $x = p/q$, define $s = p + q$. Under the operation $x \rightarrow 2x + 1 = \frac{2p+q}{q}$, $s = p + q \rightarrow 2(p + q)$ if q is odd; or $s = p + q \rightarrow p + q$ if q is even. Under the operation $x \rightarrow \frac{x}{x+2} = \frac{p}{p+2q}$, $s = p + q \rightarrow 2(p + q)$ if p is odd; or $s = p + q \rightarrow p + q$ if p is even. For the integer $z = 2008/1$, $s = 2009$ is odd, and so z is obtained by operations such that s is unchanged. So, at the beginning, the original integer $x_0 = \frac{p}{q}$ satisfies $p + q = 2009$ and $q = 1$, so $x_0 = p = 2008$.

8. Label the stones from left to right with numbers $1, 2, \ldots, 2009$.

   Suppose that the black stones from left to right are $a_1, a_2, \ldots, a_k$ at some moment, where k may vary after an *action*. Construct a variant

$$S = \sum_{i=1}^{k} (-1)^{i+1} a_i.$$

   We investigate how S changes to S' after an *action* is taken on stone a_j.

   (I) If $a_j \neq 1, 2009$. We show that $S' = S$.

   If stones $a_j - 1, a_j + 1$ are white, then they change to black after the *action*, with other stones unchanged. So,

$$S' - S = \left((-1)^{j+1}(a_j - 1) + (-1)^{j+2} a_j + (-1)^{j+3}(a_j + 1)\right)$$
$$- (-1)^{j+1} a_j = 0.$$

   If stones $a_j - 1, a_j + 1$ are in different colours, say, stone $a_j - 1$ is white, stone $a_j + 1$ is black (that is $a_{j+1} = a_j + 1$), similarly,

$$S' - S = \left((-1)^{j+1} a_j + (-1)^{j+2}(a_j + 1)\right)$$
$$- \left((-1)^{j+1}(a_j - 1) + (-1)^{j+2} a_j\right) = 0.$$

   (II) Stone a_j is the last one, that is, $j = k, a_k = 2009$. We introduce a virtual black stone $a_{k+1} = 2010$. Let

$$S_1 = \sum_{i=1}^{k+1} (-1)^{i+1} a_i = S + (-1)^{k+2} 2010.$$

   In case (I), we know that $S_1' = S_1$, stone 2010 turns white after the *action*. So,

$$S' = S_1' = S_1 = S + (-1)^{k+2} 2010.$$

   (III) Stone a_j is the first, that is, $j = 1, a_1 = 1$.

   If $a_2 = 2$ before the *action*, then

$$S' = 1 - \sum_{i=3}^{k} (-1)^{i+1} a_i = -a_1 + a_2 - \sum_{i=3}^{k} (-1)^{i+1} a_i = -S.$$

   If $a_2 > 2$ before the *action*, then

$$S' = 1 - 2 + \sum_{i=3}^{k+1} (-1)^{i+1} a_{i-1} = -a_1 + \sum_{i=2}^{k} (-1)^{i+2} a_i = -S.$$

Summing up (I), (II), and (III), we see that $S' \equiv \pm S \pmod{2010}$. If all stones are black, then $S = \sum_{i=1}^{2009} (-1)^{i+1} i = 1005 \equiv -1005 \pmod{2010}$ is an invariant. So, at the beginning, the stone 1005 is black.

Finally, we construct *actions* that turn all stones black.

Denote $m = 1005$, and suppose stones $m - k, m - k + 1, \ldots, m + k (k \in \mathbb{N})$ are black.

If $k = 2l$ $(l \in \mathbb{N})$, then take *actions* successively on the following stones:

$$m - 2l, m + 2l, m - (2l - 2), m + (2l - 2), \ldots, m - 2, m + 2, m;$$

If $k = 2l + 1$ $(l \in \mathbb{N})$, then take *actions* successively on the following stones:

$$m - (2l - 1), m + (2l - 1), m - (2l - 3), m + (2l - 3), \ldots, m - 1, m + 1.$$

Then, the black stones are on positions from $m - k - 1$ to $m + k + 1$.

For $k = 0$ to $k = 1004$, take the *actions* described above, then all stones become black.

Chapter 20

1. Let a simple graph G with n vertices represent n people. Two people u and v are friends if and only (u, v) is an edge of G.

   The degree $d(x)$ of vertex x of G satisfies $0 \leq d(x) \leq n - 1$, which means $d(x)$ people are friends of person x.

   Suppose, on the contrary, all degrees of the vertices are distinct. Then there is a vertex u with degree 0 and a vertex v with degree $n - 1$, v being adjacent to u. That contradicts $d(u) = 0$.

2. The answer is negative. Suppose, on the contrary, there is a polyhedron that has an odd number of faces and each face has an odd number of edges. Make a graph G whose vertices v's represent the faces, and (v_i, v_j) is an edge of G if and only if two faces v_i, v_j have a common edge. So, $d(v)$ is the number of sides of the face v. The sum of these odd number of odd degrees is odd. On the other hand, this sum is twice the number of sides, since each side is counted twice. A contradiction.

3. Let G be a simple graph with 100 vertices, in which each vertex represents a member of the group. Two persons x and y know each other if and only if (x, y) is an edge of G. Then, the problem is: of any four vertices of G, there is at least one vertex adjacent to the other three vertices. Let l be the number of vertices with degree 99, find the minimum m of l. We shall prove that $m = 97$.

   If G is complete, the degree of each vertex is 99, so $l = 100$.

   If G is not complete, there are two non-adjacent vertices u and v, $d(u) \leq 98, d(v) \leq 98$. So $l \leq 98$. For four vertices u, v, x, and y, x or y must be adjacent to the other three vertices, which implies that any two vertices x and y except u and v are adjacent, and so $\deg(x) \geq 97$. If vertex x is not adjacent to u and v, then $\deg(x) = 97$, and any vertex

y except u, v, and x has degree 99. So, $l = 97$ is the least possible value. That is, there are at least $m = 97$ people who know all the other people.

Remark. The problem can be generalized to any n people, and the answer is $n - 3$.

4. Let G be a simple graph with five vertices representing the five irrational numbers, respectively. Two vertices are adjacent if and only if the sum of their corresponding numbers is rational. Then the problem is to prove there are three pairwise non-adjacent vertices.

   First, we show that there is no triangle in the graph. In fact, if (x, y), (y, z), (z, x) are edges, then, $x + y, y + z, z + x$ are all rational numbers, and consequently, x, y, z are all rational numbers, a contradiction. Similarly, there is no pentagon in the graph.

   If there is a vertex x that is adjacent to at least three vertices y, z, u, then these three vertices y, z, u are pairwise non-adjacent. Otherwise, there will be a triangle with x as its vertex.

   If there is a vertex x that is not adjacent to three vertices y, z, u (they do not form a triangle), then x with two vertices of y, z, u are pairwise non-adjacent.

   Otherwise, each vertex is adjacent to exactly two vertices. So, there will be a pentagon by the Euler theorem. A contradiction.

Remark. In fact, we have just proved a graph theorem: in a red–blue colouring of complete 5-graph K^5, if there is no pentagon in red, then there is a triangle in red or blue.

5. The answer is negative.

   Make a graph G with vertices $1, 2, \ldots, 13$, in which vertices i and j are adjacent if $3 \le |i - j| \le 5$. Suppose, on the contrary, the graph G has a Hamilton cycle C (the cycle that passes each vertex just once). Divide the vertices of G into two sets:

$$A = \{1, 2, 3, 11, 12, 13\}, B = \{4, 5, 6, 7, 8, 9, 10\}.$$

   Note that the vertices in A are pairwise non-adjacent. So, each vertex of A is adjacent to two vertices of B, there are 12 edges between A and B. Since C has 13 edges, the last edge of C is incident with two vertices of B, denoted by (i, j), $i < j$.

On the other hand, vertex 4 of B is adjacent to only one vertex 1 of A, and vertex 10 of B is adjacent to only one vertex 13 of A, so, $(i, j) = (4, 10)$. But (4,10) is not an edge. A contradiction.

6. Let the n-gon with all its diagonals form a connected planar graph G, every intersection of two diagonals being a vertex of G.

Denote the order of G by V and the size of G by E, the number of faces of G by F, where the exterior of the n-gon is also a face.

Since every four vertices of the n-gon uniquely correspond to the intersect of two diagonals, the number of all intersections of the diagonals is $\binom{n}{4}$. Thus,

$$V = n + \binom{n}{4}.$$

The degree of each vertex of the n-gon is $n - 1$. The degree of the vertex of the intersection of diagonals is 4. Thus,

$$E = \frac{1}{2}\left(n(n-1) + 4\binom{n}{4}\right).$$

Substitute the above two equations into Euler's formula $V - E + F = 2$ for planar connected graphs, and yield

$$F - 1 = E - V + 1 = \frac{1}{2}\left(n(n-1) + 4\binom{n}{4}\right) - \left(n + \binom{n}{4}\right) + 1$$

$$= \frac{1}{24}(n-1)(n-2)(n^2 - 3n + 12).$$

That is, the n-gon is divided into $\frac{1}{24}(n-1)(n-2)(n^2 - 3n + 12)$ regions.

7. Consider the graph G with 20 vertices $v_1, v_2, \ldots, v_{20}$, which represent the 20 members, and (v_i, v_j) is an edge of G if and only if v_i and v_j take part in a game.

Denote the degrees of the points by $d_i, i = 1, 2, \ldots, 20$, then $d_i \geq 1$. Since there are 14 edges in G, we see that

$$d_1 + d_2 + \cdots + d_{20} = 2 \times 14 = 28.$$

Delete $d_i - 1$ edges incident with vertex v_i, $i = 1, 2, \ldots, 20$, and so at most $(d_1 - 1) + (d_2 - 1) + \cdots + (d_{20} - 1) = 8$ edges are deleted. Then, there are at least $14 - 8 = 6$ edges left, and the degree of each vertex is at most 1. So, these 6 edges are incident with 12 distinct vertices, there is a set of 6 games with 12 distinct players.

8. If these n positive integers can be divided into 41 groups $u_1, u_2, \ldots, u_{41}$ with equal sum 49, and into 49 groups $v_1, v_2, \ldots, v_{49}$ with equal sum 41, consider a graph with vertices $u_i (1 \leq i \leq 41)$ and $v_j (1 \leq j \leq 49)$, and (u_i, v_j) is an edge if and only if u_i and v_j have a common integer. We show that G is connected.

Consider the maximal connected component G' of G. Suppose that G' contains vertices $u_{i_s}, s = 1, 2, \ldots, a$ with total sum $49a$, and $v_{j_t}, t = 1, 2, \ldots, b$, with total sum $41b$. Since each x_k in some $u_{i_s} (1 \leq s \leq a)$ must also be in some $v_{j_t} (1 \leq t \leq b)$, and vice versa (otherwise G' is not maximal). We see that $49a = 41b$. So, $a + b \geq 41 + 49 = 90$. Consequently, $G' = G$, that is, G is connected. So, G has at least 89 edges, and $n \geq 89$.

On the other hand, if $n = 89$, we can check that the sequence

$$x_1 = x_2 = \cdots = x_{41} = 41, \ x_{42} = x_{43} = \cdots = x_{81} = 8,$$

$$x_{82} = x_{83} = \cdots = x_{89} = 1$$

satisfies the requirement. So, the answer is 89.

9. First note that, among 10 distinct non-zero numbers, if there exists $x \in \mathbb{Q}$, for any other number t, either $x + t$ or xt is rational, then $t \in \mathbb{Q}$.

Let K^{10} be a two-colour complete graph with vertices as the ten numbers. If $u + v$ is rational, then colour the edge (u, v) blue; otherwise, if uv is rational, colour the edge (u, v) red. By the Ramsey theorem, the graph has a monochromatic triangle.

(I) If the triangle is blue, then there are three numbers x, y, z where the sum of any two numbers is rational. So, $x, y, z \in \mathbb{Q}$, and the 10 numbers are all rational.

(II) If the triangle is red, then there are three numbers x, y, z where the product of any two of them is rational. So, $x^2 = xy \cdot xz/(yz) \in \mathbb{Q}$. Let $x = m\sqrt{a}$, where $a \in \mathbb{Q}$, $m = \pm 1$. Since $xy = m\sqrt{a}y = b \in \mathbb{Q}$, we have $y = b\sqrt{a}/(ma) = c\sqrt{a}$, where $c \in \mathbb{Q}$, $c \neq m$. For t other than x, y: (i) xt or yt is rational, then in the same way, we see that $t = d\sqrt{a}$, where $d \in \mathbb{Q}$, and hence $t^2 \in \mathbb{Q}$; (ii) $xt, yt \notin \mathbb{Q}$, then $x + t, y + t \in \mathbb{Q}$, but this is impossible, since $(x + t) - (y + t) = (m - c)\sqrt{a} \notin \mathbb{Q}$.

Summing up, we have proved that either every number is rational or the square of every number is rational.

10. Let the directed graph G have vertices as the elements of set X, and let E be the set of directed edges of G. For any $x, y \in X, x \neq y$, if $f(x, y) = 1$, the edge $(x, y) \in E$. So, for any two vertices x, y, only one of (x, y) and $(y, x) \in E$ (such a graph G is called a directed complete n-graph $\overline{K^n}$, or a *tournament*).

We prove by induction that there is a directed path $x_1 \rightarrow x_2 \rightarrow \cdots \rightarrow x_n$ of length $n - 1$ (with possible relabelling).

If $n = 2$, obviously, we can label the vertices such that $x_1 \rightarrow x_2$.

Suppose that for $n - 1 \geq 2$, there is a directed path $x_1 \rightarrow \cdots \rightarrow x_{n-1}$ of length $n - 2$.

So for $n \geq 3$, we put aside x_n and all edges incident to x_n, and by the inductive assumption, there is a directed path $x_1 \rightarrow x_2 \rightarrow \cdots \rightarrow x_{n-1}$. Now there are only two cases:

(1) $x_{n-1} \rightarrow x_n$ or $x_n \rightarrow x_1$ is an edge, then the directed path $x_1 \rightarrow x_2 \rightarrow \cdots \rightarrow x_{n-1} \rightarrow x_n$; or $x_n \rightarrow x_1 \rightarrow x_2 \rightarrow \cdots \rightarrow x_{n-1}$ has length $n - 1$, as desired.

(2) Let j be the largest index such that $x_j \rightarrow x_n$ is an edge. Then $1 \leq j \leq n - 2$, and $x_n \rightarrow x_{j+1}$ is an edge. The directed path $x_1 \rightarrow \cdots \rightarrow x_j \rightarrow x_n \rightarrow x_{j+1}(\rightarrow \cdots \rightarrow x_{n-1})$ has length $n - 1$, as desired. (a) is proved.

We have just proved that a *tournament* of order n always has a Hamilton path $x_1 \rightarrow x_2 \rightarrow \cdots \rightarrow x_n$. Next, we shall prove that if and only if condition (b) is satisfied, then there exists a Hamilton cycle: $x_1 \rightarrow \cdots \rightarrow x_n \rightarrow x_1$.

Clearly condition (b) is necessary. We now prove condition (b) is sufficient.

By condition (b), take any $x \in X$, choose partition $A = \{a \in X | x \rightarrow a\} B = \{x\} \cup \{b \in X | b \rightarrow x\}$. There are $a \in A$ and $b \in B$, such that $f(a, b) = 1$. So there exists a cycle of length 3: $x \rightarrow a \rightarrow b \rightarrow x$. So, the conclusion follows if $n = 3$.

If $n \geq 4$, let $x_1 \rightarrow x_2 \rightarrow \cdots \rightarrow x_k \rightarrow x_1$ be the longest cycle. We shall prove that $k = n$. Otherwise, if $k < n$, denote $A = \{x_1, x_2, \ldots, x_k\}$. then by condition (b), there is a pair (x_i, b), $b \in B = X \backslash A$, $x_i \in A$, such that $f(x_i, b) = 1$. Relabel the indices such that $i = 1$. Then for the partition $P = \{b \in B | f(x_1, b) = 1\}$, $Q = A \cup \{c \in B | f(c, x_1) = 1\}$, there is a pair (x_{k+1}, a), $x_{k+1} \in P$, $a \in Q$, such that $f(x_{k+1}, a) = 1$. If there exists $a = x_j \in A$. Then let $s + 1$ be the smallest index such that $f(x_{k+1}, x_{s+1}) = 1$, $s \in \{1, \ldots, j-1\}$, so, $f(x_s, x_{k+1}) = 1$. Hence, there is a longer cycle $x_1 \to \cdots \to x_s \to x_{k+1} \to x_{s+1} \to \cdots \to x_k$. Else $f(x_{k+1}, a) = 0$, for any $a \in A$, that is, $f(a, x_{k+1}) = 1$ for any $a \in A$, especially $f(x_k, x_{k+1}) = 1$. So, when $n > k+1$, there is $x_{k+2} \in Q \cap B$, such that $f(x_{k+1}, x_{k+2}) = 1$. So, $x_1 \to \cdots \to x_k \to x_{k+1} \to x_{k+2} \to x_1$ is a longer circle.

Printed in the United States
by Baker & Taylor Publisher Services